INORGANIC LAB MANUAL

(For B.Sc. and M.Sc. Students)

S. Mumtazuddin
Professor
University Deptt. of Chemistry
B.R.A. Bihar University, Muzaffarpur

Shailendra Kumar Sinha
Lecturer
P.G. Deptt. of Chemistry
L.S. College, Muzaffarpur

ATLANTIC
PUBLISHERS & DISTRIBUTORS (P) LTD

Published by
ATLANTIC
PUBLISHERS & DISTRIBUTORS (P) LTD
7/22, Ansari Road, Darya Ganj,
New Delhi-110002
Phones : +91-11-23273880, 23275880, 23280451
Fax : +91-11-23285873
Web : www.atlanticbooks.com
E-mail : info@atlanticbooks.com

Branch Office
5, Nallathambi Street, Wallajah Road,
Chennai-600002
Phones : +91-44-64611085, 32413319
E-mail : chennai@atlanticbooks.com

Printed in India at Nice Printing Press, A-33/3A, Site-IV, Industrial Area, Sahibabad, Ghaziabad, U.P.

PREFACE

This book *Inorganic Lab Manual* has been designed to meet the demands of all levels of students studying Chemistry. It is meant for the General/Subsidiary, Honours and Postgraduate students and incorporates the syllabi of all Indian universities.

The language of the book is simple and easy to follow. "How to perform the experiment stepwise and how to write the experiment" is a special attraction of this book. The experiment where calculation is needed, has been presented with expected data and required calculation so as to facilitate understanding by the students. In short, it will serve as a guide to the students.

The book includes complete theory, reasoning and reactions of each experiment. It also provides under the heading 'things to remember' some important points which students are required to remember to help them retain the basics.

This book will be helpful to the teachers and lab assistants in preparing solutions, reagents, indicators, etc. It also enlists the hazardous chemicals, their effects and precautionary and remedial measures to be taken in a lab.

We hope that this book will adequately meet the need of those students for whom it is meant. We shall be grateful to those who suggest ways for further improving this book.

S. Mumtazuddin
Shailendra Kumar Sinha

CONTENTS

1 QUALITATIVE INORGANIC ANALYSIS

1. INTRODUCTION

The term qualitative inorganic analysis means the systematic examination of salts in the given mixture on the basis of their properties. It is not necessary to identify the actual compounds present in the mixture. It is sufficient to detect radicals present in it.

When an acid reacts with a base, salt and water are formed.

Acid + Base $\rightarrow$ Salt + Water

$HCl + NaOH \rightarrow NaCl + H_2O$

It is clear from the above equation that the salt sodium chloride has two parts, one is sodium ion (Na^+) and other is chloride ion (Cl^-). Sodium ion is known as basic radical (or cation) because it has come from base, sodium hydroxide and chloride ion (or anion) is known as acid radical because it has come from acid, hydrochloric acid. Thus identification of salt means identification of its two radicals separately.

1.1. Common Basic Radicals

Group I	Hg^+,		Ag^+,		Pb^{2+}
Group IIA	Pb^{2+},	Hg^{2+},	Bi^{3+},	Cu^{2+},	Cd^{2+}
Group IIB	As^{3+},	Sn^{4+},	Sb^{3+}		
Group III	Fe^{3+},	Cr^{3+},	Al^{3+}		
Group IV	Ni^{2+},	Co^{2+},	Mn^{2+},	Zn^{2+}	

Group V Ca^{2+}, Sr^{2+}, Ba^{2+}

Group VI Mg^{2+}, Na^{+}, K^{+}, NH_4^{+}

The basic radicals mentioned above have been arranged in seven groups. Since ammonium ion is confirmed by dry test, hence it may not be included in any group.

1.2. Common Acid Radicals

CO_3^{2-}	Carbonate ion
NO_3^{-}	Nitrate ion
NO_2^{-}	Nitrite ion
SO_4^{2-}	Sulphate ion
SO_3^{2-}	Sulphite ion
S^{2-}	Sulphide ion
Cl^{-}	Chloride ion
Br^{-}	Bromide ion
I^{-}	Iodide ion

In addition to the above acid radicals following interfering radicals may also be given:

PO_4^{3-}	Phosphate ion
BO_3^{3-}	Borate ion
SiO_3^{2-}	Silicate ion
F^{-}	Fluoride ion
COO^{-} | COO^{-}	Oxalate ion
$CH(OH)COO^{-}$ | $CH(OH)COO^{-}$	Tartarate ion

The interfering radicals mentioned above are also known as non-oxidising interfering radicals. There are also a few acid radicals which are known as oxidising interfering radicals:

MnO_4^{-}	Permanganate ion
CrO_4^{2-}	Chromate ion

$Cr_2O_7^{2-}$	Dichromate ion
AsO_4^{3-}	Arsenate ion
AsO_3^{3-}	Arsenite ion

1.3. Dry Tests and Wet Tests

There are two main classes in which identification of basic radicals and acid radicals can be divided.

Dry tests are those tests which are directly applied to the given salt mixture and wet tests are those tests which are done by making a solution of salt mixture in a suitable solvent. Dry tests are preliminary and wet tests are confirmatory for acid and basic radicals.

The dry tests occupy a very significant place in the detection of radicals and should never be under-rated. They take very little time to perform and often give very valuable information.

Here students are advised to perform the identification of acid radicals first. It is so because if oxidising interfering radical be present in the mixture, it must be removed before separation of group II and if non-oxidising interfering radical be present, it must be removed before separation of III. Thus the knowledge of acid radicals in the supplied mixture is essential before the identification of basic radicals. Finally, we would like to advise the students to perform the analysis of salt mixture in following order:

1. Physical characteristics
2. Dry (preliminary) tests for acid radicals
3. Wet (confirmatory) tests for acid radicals
4. Dry (preliminary) tests for basic radicals
5. Wet (confirmatory) tests for basic radicals

2. THEORETICAL BASIS OF QUALITATIVE INORGANIC ANALYSIS

For identification of acid radicals present in the salt mixture, individual test is done for individual acid radical. But identification of basic radicals is very systematic for which one should have some theoretical ideas.

2.1. Solubility Product

The solubility product of a sparingly soluble salt is the product of the concentrations of its constituent ions (cation and anion) in its saturated solution, expressed as gram-ion per litre, each concentration term being raised to a power representing the number of those ions formed by the dissociation of the molecule of the salt.

In other words, solubility product of a sparingly soluble salt is its ionic product in its saturated solution, e.g.,

$$B_xA_y \rightleftharpoons xB^+ + yA^-$$

$$\therefore \quad S_{B_xA_y} = [B^+]^x \times [A^-]^y$$

where $S_{B_xA_y}$ represents solubility product of salt B_xA_y.

2.2. Principle of Solubility Product

1. When the ionic product of a sparingly soluble salt in a solution is less than its solubility product, the salt remains dissolved in the solution.

2. When the ionic product of a sparingly soluble salt in a solution exceeds its solubility product, the salt precipitates out from the solution.

3. When the ionic product of a sparingly soluble salt in a solution is equal to its solubility product, the salt precipitates out as well as dissolves in solution. Actually this case arises in saturated solution.

2.3. Common Ion Effect

The suppression in ionisation of a weak electrolyte caused by addition of a strong electrolyte having an ion common to the weak electrolyte is known as common ion effect, e.g.,

$$\underset{\text{(Strong electrolyte)}}{NH_4Cl} \rightleftharpoons NH_4^+ + Cl^-$$

$$\underset{\text{(Weak electrolyte)}}{NH_4OH} \rightleftharpoons NH_4^+ + OH^-$$

$$K_{NH_4OH} = \frac{[NH_4^+][OH^-]}{[NH_4OH]}$$

$$= 1.8 \times 10^{-5}$$

On addition of ammonium chloride, concentration of NH_4^+ ion increases in the solution. So, in order to maintain the constant value of K_{NH_4OH}, the dissociation of ammonium hydroxide is suppressed. Due to this suppression, $[NH_4OH]$ increases and $[OH^-]$ decreases.

3. PHYSICAL CHARACTERISTICS

State	—	Crystalline/Amorphous
Colour	—	....
Solubility	—	Soluble/Insoluble in water

State and colour of the mixture should be carefully observed and noted, because sometimes they give valuable information, but it will not be advisable to place too much reliance on it. In a mixture of inorganic compounds, however, colour does not give much information regarding the nature of compounds present as colour of one of the compounds may be masked by that of the other.

White or Colourless:

Salts of Na, K, NH_4, Mg, Ca, Sr, Ba, Bi, Zn and Al are usually white or colourless unless the acid radical (e.g., chromate, permanganate, etc.) is coloured.

Blue:

Anhydrous cobalt salts, hydrated cupric salts.

Green:

$CuCl_2 \cdot 2H_2O$, $CuCO_3$, salts of nickel and chromium.

Light Green:

Ferrous salts except FeS.

Light Pink:

Hydrated salts of Mn.

Dark Pink:

Cobalt salts.

Yellow to Brown:

Ferric salts

Black:

CuO, MnO_2, PbS, CuS, HgS, FeS, CoS, NiS

Yellow:

Chromates, ferric ion, CdS, As_2S_3, SnS_2, PbI_2.

Orange-red:

Dichromates.

Purple:

Permanganates.

4. DRY TESTS FOR ACID RADICALS

Dry tests are those tests which are directly applied to the given salt mixture.

The identification of acid radicals does not follow a well organised plan as is the case with basic radicals. They cannot be easily grouped on any systematic basis.

In dry tests, the acid radicals are generally identified by the characteristics of the gases obtained when a salt is allowed to react with mineral acids. The gases so obtained are identified by their odour and colour. After having suspected the nature of the gas by its colour and odour, the gas is passed with the help of a bent delivery tube (as shown below) into a suitable reagent. Colour-change and precipitate formed is noted.

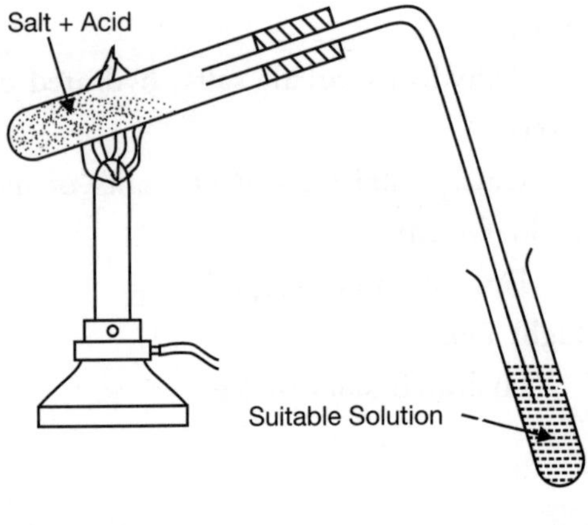

To perform the dry tests for acid radicals, following two tests can be done:

1. Dilute hydrochloric acid test and
2. Concentrated sulphuric acid test

These tests have been comprised in Table 1 given below. Students are advised to follow the table only inside the laboratory and they should go through explanation of the several tests, mentioned after the table, at home.

Table 1

Experiment	Observation	Inference
1. Dil. hydrochloric acid test: A little amount of substance is taken in a test tube. 2-3 ml. dil. HCl is added to it and then heated towards the end of the reaction.	(i) Brisk effervescence observed in cold; colourless and odourless gas is evolved. This gas is passed through lime water. Lime water becomes milky.	May be CO_3^{2-} or HCO_3^-
	(ii) Slow effervescence is found in cold. Colourless gas with smell of rotten egg evolves. This gas is passed through acidified lead acetate solution. The solution becomes black.	May be S^{2-} (Sulphides of certain metals give this test only upon warming it with dil. HCl and granulated zinc)
	(iii) Slow effervescence is found in cold. Colourless gas with smell of burning sulphur evovles. It is passed through acidified potassium permanganate solution or acidified potassium dichromate solution. The gas turns acidified potassium permanganate solution colourless or acidified potassium dichromate solution green.	May be SO_3^{2-}
	(iv) Slow effervescence is found in cold. Brown gas with characteristic smell of	May be NO_2^-

(Contd...)

Experiment		Observation	Inference
		oxides of nitrogen evolves. It is passed through freshly prepared $FeSO_4$ solution. The solution becomes brownish black. (The gas also turns starch iodide paper blue.)	
	(v)	Iodine is obtained as residue accompanied with the evolution of slight NO_2 gas.	May be I^- and NO_2^-
2. Conc. sulphuric acid test: A little amount of the substance is taken in a dry test tube. 2-3 ml. conc. H_2SO_4 is added to it. Now it is warmed gently.	(i)	Colourless gas with pungent smell evolves. It gives white dense fumes when a glass rod dipped in liquor ammonia is brought near the rim of test tube. The gas also turns moist blue litmus paper red. On adding a pinch of MnO_2 to the hot reaction mixture, greenish yellow gas with characteristic smell of chlorine evolves which bleaches wet litmus paper.	May be Cl^-
	(ii)	Oily appearance of the test tube is found in cold. On warming, colourless pungent gas evolves which corrodes the glass, if moistened glass rod is introduced into the vapour, gelatinous precipitate of silicic acid is deposited upon it.	May be F^-
	(iii)	Reddish brown gas with characteristic smell of bromine is observed. On adding a pinch of MnO_2 to the hot reaction mixture, more reddish brown (colour deepens) fumes of bromine evolves	May be Br^-

(Contd...)

Experiment	Observation	Inference
	which turns starch paper (prepared by soaking a piece of filter paper with starch solution) orange yellow.	
	(iv) Violet vapour with characteristic smell of iodine evolves which turns starch paper blue.	May be I^-
	(v) Reddish brown fumes of chromyl chloride evolves (similar in colour to bromine). It may be tested by passing into NaOH solution. NaOH solution becomes yellow. This solution is acidified with acetic acid and then treated with lead acetate solution. Consequently, yellow precipitate is formed.	May be Cl^- in presence of $Cr_2O_7^{2-}$ or CrO_4^{2-}
	(vi) Yellowish-brown gas with pungent and nitrous smell evolves. On adding a few copper turning to the reaction mixture, more of the brown gas (colour deepens) with nitrous smell evolves, turns $FeSO_4$ paper (prepared by soaking a piece of filter paper with freshly prepared $FeSO_4$ solution) brownish black.	May be NO_3^- or/and NO_2^-
	(vii) Greenish-yellow gas with suffocating smell evolves which bleaches moist litmus paper.	May be Cl^- in presence of oxidising agent (e.g., MnO_2, MnO_4^-, etc., not CrO_4^{2-} or $C_2O_7^{2-}$).

(*Contd...*)

Experiment	Observation	Inference
	(viii) Colourless gas evolves. It turns lime water milky and also burns with blue flame; no blackening.	May be COO^- | COO^-
	(ix) Colourless gas evolves. It burns with blue flame and turns lime water milky. The residue in test tube chars rapidly and odour of burnt sugar is observed.	May be $CH(OH)COO^-$ | $CH(OH)COO^-$

4.1. Explanation of Dilute Hydrochloric Acid Test

A little amount of substance is taken in a test tube. About 5 ml. dil. HCl is added to it and observed whether any reaction takes place in the cold or not. The reaction is indicated by appearance of the gas bubbles, i.e., effervescence. If a gas evolves, the colour and odour of the gas is noted. Having noted these, the content is heated to boiling and the gas passed in a suitable reagent for confirmation.

Dilute hydrochloric acid decomposes salts of volatile acids. Thus,

(i) Dilute hydrochloric acid reacts with carbonate and bicarbonate to give carbon dioxide gas. Due to evolution of CO_2 gas, effervescence takes place. CO_2 gas is colourless and odourless.

$$CO_3^{2-} + 2H^+ \rightarrow H_2O + CO_2 \uparrow$$

$$HCO_3^- + H^+ \rightarrow H_2O + CO_2 \uparrow$$

If CO_2 gas is suspected, it is passed through lime water which becomes milky due to formation of white and insoluble calcium carbonate.

$$CO_2 + Ca(OH)_2 \rightarrow \underset{\text{(White ppt.)}}{CaCO_3} + H_2O$$

However, in the excess CO_2, milky appearance disappears due to formation of soluble calcium bicarbonate.

$$CaCO_3 + H_2O + \underset{\text{(excess)}}{CO_2} \rightarrow \underset{\text{(soluble)}}{Ca(HCO_3)_2}$$

(ii) Dilute hydrochloric acid reacts with sulphide to give hydrogen sulphide gas.

$$S^{2-} + 2H^+ \rightarrow H_2S \uparrow$$

Due to evolution of H_2S gas, slow effervescence takes place. It is colourless and has a smell of rotten egg. When H_2S gas is suspected, it is passed into lead acetate solution having been acidified with acetic acid. H_2S gas reacts with lead acetate to produce black precipitate of lead sulphide.

$$H_2S + (CH_3COO)_2\ Pb \rightarrow 2CH_3COOH + \underset{\text{(Black ppt.)}}{PbS \downarrow}$$

Note: Sulphides of certain metals, e.g., As, Sb and Hg do not usually react with dil. acid to liberate H_2S. Some H_2S is, however, evolved upon warming a mixture of the sulphide, dil. HCl and granulated zinc.

(iii) Dilute hydrochloric acid reacts with sulphite to give sulphur dioxide gas.

$$SO_3^{2-} + 2H^+ \rightarrow H_2O + SO_2 \uparrow$$

Due to evolution of sulphur dioxide gas, slow effervescence takes place in cold. It is colourless and has a smell of burning sulphur. When SO_2 gas is suspected, it is passed into potassium permanganate solution acidified with dil. H_2SO_4 or potassium dichromate solution acidified with dil. H_2SO_4. SO_2 gas turns acidified potassium permanganate solution colourless and acidified potassium dichromate green.

$$[MnO_4^- + 8H^+ + 5\cancel{e} \rightarrow Mn^{2+} + 4H_2O] \times 2$$

$$[SO_2 + 2H_2O \rightarrow SO_4^{2-} + 4H^+ + 2\cancel{e}] \times 5$$

$$2MnO_4^- + 5SO_2 + 2H_2O \rightarrow \underset{\text{(Colourless)}}{2Mn^{2+}} + 5SO_4^{2-} + 4H^+$$

$$Cr_2O_7^{2-} + 14H^+ + 6\cancel{e} \rightarrow 2Cr^{3+} + 7H_2O$$

$$[SO_2 + 2H_2O \rightarrow SO_4^{2-} + 4H^+ + 2\cancel{e}] \times 3$$

$$Cr_2O_7^{2-} + 2H^+ + 3SO_2 \rightarrow \underset{\text{(Green)}}{2Cr^{3+}} + 3SO_4^{2-} + H_2O$$

Acidified potassium permanganate becomes colourless due to formation of Mn^{2+} ion and acidified potassium dichromate becomes green due to formation of Cr^{3+} ion.

(iv) Dilute hydrochloric acid reacts with nitrite and gives nitrous acid first.

$$NO_2^- + H^+ \rightarrow HNO_2$$

Nitrous acid is unstable, hence it decomposes to produce nitric acid and nitric oxide.

$$3HNO_2 \rightarrow HNO_3 + H_2O + \underset{\text{(Colourless)}}{2NO \uparrow}$$

$$4HNO_2 \rightarrow 2H_2O + \underset{\text{(Colourless)}}{4NO} + O_2$$

Nitric oxide comes in contact with air to form brown coloured nitrogen peroxide gas.

$$NO + \underset{\text{(Air)}}{\frac{1}{2}O_2} \rightarrow \underset{\text{(Brown)}}{NO_2}$$

When nitrogen peroxide gas is passed into freshly prepared ferrous sulphate solution, the solution becomes brown due to formation of complex mentioned below.

$$Fe^{2+} + NO + SO_4^{2-} \rightarrow \underset{\substack{\text{Nitrosonium iron (I) sulphate} \\ \text{(Brown)}}}{[Fe(NO)]SO_4}$$

NO_2 oxidises KI to liberate I_2. The liberated I_2 forms a deep blue complex with starch.

$$NO_2 + 2KI + H_2SO_4 \rightarrow K_2SO_4 + I_2 + H_2O + NO$$

(v) If both nitrite and iodide are present in the salt mixture, dilute hydrochloric acid reacts with them to form I_2 and NO. Iodine is obtained in the test tube as residue and NO reacts with air to evolve brown coloured nitrogen peroxide gas.

$$\left[NO_2^- + 2H^+ + \cancel{e} \rightarrow NO + H_2O\right] \times 2$$

$$2I^- \rightarrow I_2 + 2\cancel{e}$$

$$\overline{2NO_2^- + 2I^- + 4H^+ \rightarrow I_2 + 2H_2O + 2NO}$$

$$NO + \underset{\text{(Air)}}{\frac{1}{2}O_2} \rightarrow \underset{\text{(Brown)}}{NO_2}$$

Things to Remember

(i) Dilute H_2SO_4 can also be used in place of dilute HCl, but dil. HCl should be preferred because H_2SO_4 forms a crust of insoluble sulphates in the case of alkaline earth (Ca, Sr, Ba, etc.) carbonates.

(ii) In dilute hydrochloric acid test, gases are often evolved quickly. Therefore, tests for gases should be performed at once after the addition of the dilute hydrochloric acid to the mixture.

(iii) Boiling of the contents should be avoided, otherwise the vapour of HCl evolved may make the recognition of other gases difficult.

(iv) SO_2 also turns lime water milky, so presence of CO_2 should be reported only if the gas evolved is odourless.

(v) Nitrite ion may be confirmed by dilute hydrochloric acid test.

4.2. Explanation of Concentrated Sulphuric Acid Test

A little amount of solid mixture is taken in a dry test tube. 2-3 ml. concentrated sulphuric acid is added to it carefully. If there is effervescence in cold, it may be due to any radical indicated with dilute hydrochloric acid. In such a case, reaction mixture is shaken first and warmed gently to decompose the previously indicated radicals. Then the test tube is heated gently, the mouth of which should be inclined away from the observer, and the gas evolved is detected by odour and holding the suitable paper (prepared by moistening a piece of filter paper with the suitable reagent) near the test tube without touching it.

A few nitrates give brown fumes only when strongly heated. So, towards the end of reaction, contents should be strongly heated. Only then nitrate should be declared present or absent.

Carbonate, sulphide, sulphite and nitrite react with concentrated sulphuric acid in similar fashion as with dilute hydrochloric acid but reaction is more vigorous with conc. H_2SO_4. So all the observations noted under the dilute hydrochloric acid test are repeated with conc. H_2SO_4. Hence those observations are omitted here.

(i) Concentrated sulphuric acid reacts with chloride to liberate colourless and pungent smelling hydrogen chloride gas.

$$Cl^- + H^+ \rightarrow HCl \uparrow$$

But $HgCl_2$ is an exception because on account of its volatility, it sublimes too easily and gives sublimate of mercuric chloride. In this case, the presence of chloride is indicated by heating with MnO_2 and conc. H_2SO_4 when chlorine is evolved.

When a glass rod wet with liquor ammonia is brought near the rim of the test tube, we get white dense fumes of ammonium chloride.

$$NH_3 + HCl \rightarrow NH_4Cl$$

This is a method to test hydrogen chloride gas evolved. HCl gas also turns wet blue litmus paper red. This is another test of HCl gas.

On adding a pinch of MnO_2 to the hot reaction mixture, greenish yellow gas with characteristic smell of chlorine evolves,

$$MnO_2 + 4H^+ + 2\not{e} \rightarrow Mn^{2+} + 2H_2O$$

$$2Cl^- \rightarrow Cl_2 + 2\not{e}$$

$$\overline{MnO_2 + 2Cl^- + 4H^+ \rightarrow Mn^{2+} + 2H_2O + Cl_2}$$

The chlorine gas so evolved bleaches wet litmus paper. This is the test of chlorine as we know that chlorine has bleaching action.

(ii) Conc. H_2SO_4 reacts with fluoride to liberate colourless, pungent smelling and corrosive hydrogen fluoride gas.

$$F^- + H^+ \rightarrow HF$$

The test tube acquires a greasy appearance as a result of the corrosive action of the vapour on the silica of the glass, which liberates the gas, silicon tetrafluoride, SiF_4.

$$SiO_2 + 4HF \rightarrow SiF_4 \uparrow + 2H_2O$$

By holding a moistened glass rod in the vapour, a gelatinous silicic acid, H_4SiO_4, is deposited on the rod, which is formed as a result of the following reaction:

$$3SiF_4 + 4H_2O \rightarrow 2H_2SiF_6 + H_4SiO_4$$

(iii) If concentrated sulphuric acid is poured on some solid bromide, first a reddish-brown solution is formed, later reddish-brown vapours of bromine accompany the hydrogen bromide (fuming in moist air) which is evolved.

$$Br^- + H^+ \rightarrow HBr$$

$$SO_4^{2-} + 4H^+ + 2\cancel{e} \rightarrow SO_2 + 2H_2O$$

$$2Br^- \rightarrow Br_2 + 2\cancel{e}$$

$$\overline{SO_4^{2-} + 4H^+ + 2Br^- \rightarrow Br_2 + 2H_2O + SO_2}$$

On addition of a pinch of manganese dioxide, evolution of Br_2 is accelerated.

$$MnO_2 + 4H^+ + 2\cancel{e} \rightarrow Mn^{2+} + 2H_2O$$

$$2Br^- \rightarrow Br_2 + 2\cancel{e}$$

$$\overline{MnO_2 + 2Br^- + 4H^+ \rightarrow Mn^{2+} + 2H_2O + \underset{\text{(Reddish-brown)}}{Br_2}}$$

If conc. H_3PO_4 is taken instead of conc. H_2SO_4, only HBr is formed.

$$Br^- + H_3PO_4 \rightarrow HBr + H_2PO_4^-$$

(iv) Conc. H_2SO_4 reacts with solid iodide to form HI gas. It is further oxidised by conc. H_2SO_4 into violet vapour of iodine

$$I^- + H^+ \rightarrow HI$$

$$SO_4^{2-} + 4H^+ + 2\cancel{e} \rightarrow SO_2 + 2H_2O$$

$$2I^- \rightarrow I_2 + \cancel{2e}$$

$$\overline{SO_4^{2-} + 2I^- + 4H^+ \rightarrow \underset{\text{(Violet vap.)}}{I_2} + 2H_2O + SO_2}$$

Iodine so evolved is tested by starch paper. Iodine turns starch paper blue.

If conc. H_3PO_4 is taken instead of conc. H_2SO_4, only HI is formed.

$$I^- + H_3PO_4 \rightarrow HI + H_2PO_4^-$$

(v) If both chloride and chromate or dichromate are present in salt mixture, conc. H_2SO_4 reacts with them to evolve reddish-brown fumes of chromyl chloride (similar in colour to bromine).

$$2CrO_4^{2-} + \underset{\text{(from conc. } H_2SO_4)}{2H^+} \rightarrow Cr_2O_7^{2-} + H_2O$$

$$4Cl^- + Cr_2O_7^{2-} + 6H^+ \rightarrow 2CrO_2Cl_2 + 3H_2O$$

When chromyl chloride gas is passed into sodium hydroxide solution, the solution becomes yellow.

$$CrO_2Cl_2 + 4OH^- \rightarrow \underset{\substack{\text{chromate ion} \\ \text{(yellow)}}}{CrO_4^{2-}} + 2Cl^- + 2H_2O$$

Yellow colour of the solution is due to formation of chromate ion. The resulting solution is now acidified with acetic acid and then lead acetate solution is added. Consequently, yellow precipitate is obtained.

$$CrO_4^{2-} + Pb^{2+} \rightarrow \underset{\text{(Yellow ppt.)}}{PbCrO_4}$$

(vi) Conc. H_2SO_4 reacts with nitrate to give yellowish-brown vapour of HNO_3.

$$NO_3^- + H^+ \rightarrow \underset{\text{(Yellowish-brown vapour)}}{HNO_3}$$

Brown fumes of NO_2 are also liberated by the decomposition of HNO_3.

$$4HNO_3 \rightarrow 2H_2O + \underset{\text{(Brown)}}{4NO_2} + O_2$$

Nitric acid is also decomposed to copiously evolve NO_2 (brown) by strong heating with metallic copper.

$$\left[NO_3^- + 2H^+ + \cancel{e} \rightarrow NO_2 + H_2O\right] \times 2$$

$$Cu \rightarrow Cu^{2+} + 2\cancel{e}$$

$$Cu + 2NO_3^- + 4H^+ \rightarrow Cu^{2+} + 2NO_2 + 2H_2O$$

Nitrite gives HNO_2 by dil. HCl as well as conc. H_2SO_4.

$NO_2^- + H^+ \rightarrow HNO_2$

The HNO_2 evolved suffers auto-oxidation and is converted into HNO_3 (yellowish-brown) and NO (colourless)

$$\underset{\text{(unstable)}}{3\,HNO_2} \xrightarrow{\text{auto oxidation}} \underset{\text{(yellowish brown)}}{HNO_3} + H_2O + 2NO$$

Brown fume is due to the conversion of HNO_3 and NO into NO_2.

$$\underset{\text{(colourless)}}{NO} + \frac{1}{2}\underset{\text{(air)}}{O_2} \rightarrow \underset{\text{(brown)}}{NO_2}$$

In short, nitrite gives brown fumes with dilute HCl as well as concentrated H_2SO_4 whereas nitrate gives brown fumes when heated with concentrated H_2SO_4 only.

(vii) If both chloride and an oxidising agent (e.g., MnO_2, permanganates, etc.) are present, conc. H_2SO_4 reacts with them to produce Cl_2 gas (greenish-yellow gas with suffocating smell). This gas bleaches moist litmus paper.

$Cl^- + H^+ \rightarrow HCl$

$$MnO_2 + 4H^+ + 2\cancel{e} \rightarrow Mn^{2+} + 2H_2O$$

$$2Cl^- \rightarrow Cl_2 + 2\cancel{e}$$

$$\underset{\text{(Oxidising agent)}}{MnO_2} + 2Cl^- + 4H^+ \rightarrow Mn^{2+} + 2H_2O + Cl_2 \uparrow$$

(viii) Conc. H_2SO_4 reacts with oxalate to form CO and CO_2 gases. In this reaction conc. H_2SO_4 acts as dehydrating agent.

$$\begin{matrix} COO^- \\ | \\ COO^- \end{matrix} + 2H^+ \rightarrow \underset{\text{(Oxalic acid)}}{\begin{matrix} COOH \\ | \\ COOH \end{matrix}}$$

$$\begin{matrix} COOH \\ | \\ COOH \end{matrix} \xrightarrow{\text{conc. } H_2SO_4} H_2O + CO_2 + CO$$

The carbon dioxide gas converts lime water milky due to formation of insoluble white calcium carbonate.

$$Ca(OH)_2 + CO_2 \rightarrow \underset{\text{(white ppt.)}}{CaCO_3} + H_2O$$

The carbon monoxide burns with blue flame due to its oxidation.

$$CO + \frac{1}{2}\underset{\text{(Air)}}{O_2} \rightarrow CO_2$$

(ix) Conc. H_2SO_4 reacts with tartarate to produce carbon monoxide, carbon dioxide and carbon.

$$\begin{matrix} CH(OH)COO^- \\ | \\ CH(OH)COO^- \end{matrix} + 2H^+ \rightarrow \underset{\text{(Tartaric acid)}}{\begin{matrix} CH(OH)COOH \\ | \\ CH(OH)COOH \end{matrix}}$$

$$\underset{\text{(Tartaric acid)}}{\begin{matrix} CH(OH)\ COOH \\ | \\ CH(OH)\ COOH \end{matrix}} \xrightarrow{\text{conc. } H_2SO_4} 3H_2O + CO_2 + CO + \underset{\text{(black)}}{2C}$$

The carbon dioxide turns lime water milky, the carbon monoxide burns with blue flame and the reaction takes place with charring due to decomposition of tartaric acid into charcoal.

Things to Remember

(i) Liberated gas is not passed into suitable reagent with the help of bent delivery tube as is done in the case of dilute hydrochloric acid test, there might be back suction causing explosion.

(ii) Hot sulphuric acid is never thrown in the sink. First it is allowed to cool and then thrown away, the sink is washed with sufficient water.

(iii) Odour of any gas evolved by the action of concentrated sulphuric acid on the mixture must be noted when content is slightly warm. Odour must not be noted when the content is boiling because pungent smelling SO_2 formed due to the decomposition of H_2SO_4 will mask the odour of all other gases.

$$H_2SO_4 \rightarrow H_2O + SO_2 + \frac{1}{2}O_2$$

(iv) Permanganate dissolves in conc. H_2SO_4 forming manganese heptoxide, Mn_2O_7 and the solution becomes green.

$$2MnO_4^- + 2H^+ \rightarrow \underset{\text{(green)}}{Mn_2O_7} + H_2O$$

The solution is liable to explode spontaneously at ordinary temperature and vigorous explosion may result on warming.

(v) Dilute acid has no action on nitrates. In contrast nitrites react with dil. acids to liberate NO_2.

(vi) Oxalate and tartarate may be confirmed by conc. H_2SO_4 test.

(vii) Fluorides may be confirmed by conc. H_2SO_4 test.

5. WET TESTS FOR ACID RADICALS

Wet tests are those tests which are applied on the solution of salt mixture.

Owing to numerous interferences, the wet tests for acid radicals cannot be reduced to a systematic routine like the one for basic radicals. Therefore, we have to depend much on the dry tests. The "wet tests" for acid radicals, however, have their own significance since the radicals already detected by dry tests are looked for again and confirmed while the others, for example, sulphate, etc. are identified independently by wet tests only.

Besides general acid radicals, there are some special radicals, which are known as interfering radicals which will be discussed first.

5.1. Interfering Radicals

"Interfering radicals are those acid radicals which interfere in general course of group separation for basic radicals." Interfering radicals can be classified as depicted below:

Interfering radicals

Oxidising interfering radicals		Non-oxidising interfering radicals	
e.g., MnO_4^-	Permanganate	e.g., PO_4^{3-}	Phosphate
$Cr_2O_7^{2-}$	Dichromate	BO_3^{3-}	Borate
CrO_4^{2-}	Chromate	SiO_3^{2-}	Silicate
AsO_4^{3-}	Arsenate	F^-	Fluoride
AsO_3^{3-}	Arsenite	$\begin{array}{l} COO^- \\ \vert \\ COO^- \end{array}$	Oxalate
		$\begin{array}{l} CH(OH)\,COO^- \\ \vert \\ CH(OH)\,COO^- \end{array}$	Tartarate

Oxidising interfering radicals like permanganate, dichromate, etc. interfere in group II of group separation for basic radicals.

Suppose that metal ions of group II are not present and oxidising ion is present in the mixture. In this condition, when H_2S gas is passed through the solution acidified with HCl for separation of group II, yellow colloidal precipitate of sulphur is produced. This makes confusion with precipitate of group II.

$$Cr_2O_7^{2-} + 14H^+ + 6\cancel{e} \rightarrow 2Cr^{3+} + 7H_2O$$

$$\left[S^{2-} \rightarrow S + 2\cancel{e}\right] \times 3$$

$$Cr_2O_7^{2-} + 3S^{2-} + 14H^+ \rightarrow 2Cr^{3+} + \underset{\text{(yellow ppt.)}}{3S} + 7H_2O$$

Non-oxidising interfering radicals like phosphate, borate, etc., interfere in group III of group separation for basic radicals. In group separation, medium remains acidic upto group II. As soon as the medium changes from acidic to alkaline in group III, interference by non-oxidising interfering radicals takes place. Phosphates borates, silicates, fluorides, oxalates, tartarates, etc. of metal ions of group III to VI (except, Na^+, K^+ and NH_4^+) are insoluble in alkaline medium. Suppose, phosphate and Zn^{2+} are present in the salt mixture and metal ions of group III (Fe^{3+}, Al^{3+}, Cr^{3+}) are absent. As the solution is made alkaline for separation of group III in group separation, zinc phosphate precipitates out. This precipitate of zinc phosphate makes confusion with the precipitate of group III.

In this way, we see that oxidising interfering radicals interfere in group II and non-oxidising interfering radicals interfere in group III of group separation. So it is necessary to remove interfering radicals (if present) at proper place in group separation.

Due to interference by some acid radicals in general course of group separation for basic radicals, it is necessary to identify all the acid radicals in the mixture first. This is why we should start with identification of acid radicals.

To perform wet tests for acid radicals, it is necessary to prepare solution of the mixture.

5.2. Preparation of Solution for Wet Tests for Acid Radicals

A solution containing all acid radicals present in the given mixture, but free from heavy metals, is the first necessity for the wet tests. It is prepared as follows:

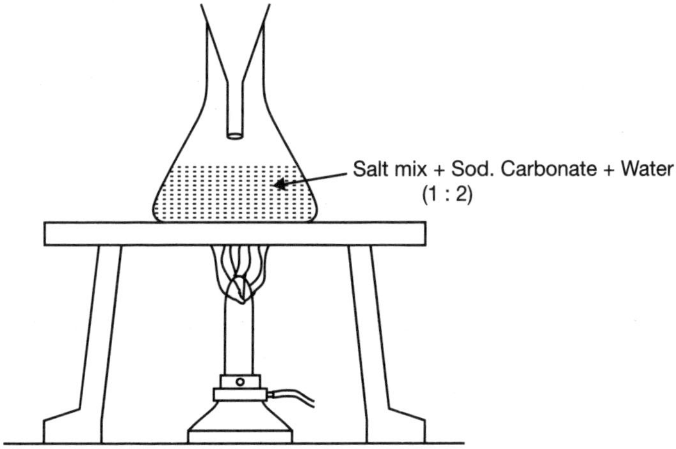

(i) Treatment of water soluble mixtures

If it is found that the given mixture is soluble in water, 5 c.c. aqueous solution is taken and equal volume of strong solution of a sodium carbonate is added in it. In case no precipitate is formed with sodium carbonate, the aqueous solution as such can be used for the wet tests, since it is free from all non-alkali metals. If a precipitate is formed with sodium carbonate, about 1 g of solid mixture and about 2 g of solid sodium carbonate are taken in a conical flask. About 40 c.c. water is added in it and boiled for about 15 minutes with a small funnel in the mouth of the conical flask to reduce the loss by evaporation. It is cooled and filtered. The filtrate is called sodium carbonate extract.

(ii) Treatment of water insoluble mixtures

If the mixture is found insoluble in water, about 1 g of the mixture and about 2 g of solid sodium carbonate are taken in a conical flask. About 40 c.c. water is added in it. It is boiled for about 15 minutes with a small funnel in the mouth of the conical flask to reduce the loss by evaporation. It is cooled and filtered. The filtrate is called sodium carbonate extract.

When salt mixture is boiled with sodium carbonate, double decomposition takes place. Suppose the given unknown substance is the salt of bivalent metal, M and an acid HA. In such case, the salt will react with sodium carbonate as under:

$$MA_2 + Na_2CO_3 \rightarrow MCO_3 + 2NaA$$

The carbonate, MCO_3 is insoluble (except carbonates of alkali metals) and sodium salt of acid radical is soluble. Thus, when the resulting mixture is filtered, sodium salt of acid radical, which is to be tested, is obtained as filtrate. This filtrate is known as sodium carbonate extract which contains all the acid radicals to be tested.

Now to perform the wet tests for acid radicals, students are advised to follow the Table 2 only inside the laboratory and they should go through explanation of the several tests, mentioned after the table, at home.

Table 2

Experiment	Observation	Inference
1. Barium chloride test:		
About 5 c.c. sodium carbonate extract is taken in a test tube. It is acidified with dil. HCl, tested with blue litmus which turns red. It is boiled to expel all the carbon dioxide. Now barium chloride solution is added.	White ppt.	SO_4^{2-} present
2. Silver nitrate test:		
About 5 c.c. sodium carbonate extract is taken in a test tube. It is acidified with dil. nitric acid, tested with blue litmus paper which turns red. It is boiled to expel all the carbon dioxide. Now silver nitrate solution is added.	(i) White curdy ppt., readily soluble in dil. NH_4OH	Cl^- present
	(ii) Pale yellow ppt., not soluble in dil. NH_4OH but soluble in liquor ammonia.	Br^- present
	(iii) Yellow ppt., insoluble even in liquor ammonia.	I^- present
	(iv) Brown or black ppt.	S^{2-} present
3. Chromylchloride test:		
Solid salt mixture and solid potassium dichromate (1:3) are taken in a test tube. About 5 c.c. conc. H_2SO_4 is added to it and heated. The evolved (reddish-brown) vapour is passed through sodium hydroxide solution. Now the solution is acidified with dilute acetic acid and then lead acetate solution is added to it.	Yellow ppt.	Cl^- present

(*Contd...*)

Experiment	Observation	Inference
4. Test of bromide:		
About 5 c.c. sodium carbonate extract is taken in a test tube. It is acidified with dil. HCl, tested with blue litmus paper which turns red. It is boiled to expel all CO_2 gas. Now carbon disulphide (or carbon tetrachloride or chloroform) and chlorine water are added to it and shaken.	Yellow or orange colouration in organic layer observed.	Br^- present
5. Test of iodide:		
About 5 c.c. sodium carbonate extract is taken in a test tube. It is acidified with dil. HCl, tested with blue litmus paper which turns red. The solution is boiled to expel all CO_2 gas. Now carbon disulphide (or carbon tetrachloride or chloroform) and chlorine water are added and shaken.	Violet colouration in organic layer observed.	I^- present
6. Test of bromide and iodide in presence of each other:		
The extract is acidified with dil. HCl. It is boiled to expel all the carbon dioxide. Chloroform (or carbon tetrachloride) is added to it. 1-2 drops of chlorine water are added and shaken.	Violet colouration in organic layer	I^- present
Addition of chlorine water is continued drop by drop until the violet colour	Reddish-brown colouration in organic layer	Br^- present

(*Contd...*)

Experiment	Observation	Inference
disappears. After the disappearance of violet colour, 1-2 drops of chlorine water are further added.		
7. Ring test:		
Sodium carbonate extract (2 c.c.) of the substance is taken in a test tube and acidified with dil. H_2SO_4. 2 c.c. freshly prepared $FeSO_4$ solution is added and shaken well.	The solution turns brown.	NO_2^- present
If the solution does not turn brown:		
To the above solution, 2-3 c.c. conc. H_2SO_4 is added by allowing it to flow down the side of the test tube, held in an inclined position.	Brown ring is formed at the junction of the aqueous layer and conc. H_2SO_4.	NO_3^- present
8. Test of nitrate in presence of bromide and iodide:		
The extract is boiled with sodium hydroxide solution till ammonium salts, if present, are completely decomposed. The mixture is then cooled and treated with Devarda's alloy or zinc dust and warmed.	Evolution of ammonia gas.	NO_3^- present
9. Test of nitrate in presence of nitrite:		
To decompose nitrite, solid salt mixture and solid ammonium chloride in the ratio of 1:3 are taken. A little water is added and boiled till	Brown ring at the junction of conc. H_2SO_4 and aqueous phase.	NO_3^- present

(Contd...)

Experiment	Observation	Inference
effervescence ceases. The residue is treated with water and filtered. Freshly prepared $FeSO_4$ solution is added to this filtrate. Now conc. H_2SO_4 is poured cautiously down the side of the test tube containing the solution.		
10. Test of sulphite:		
About 5 ml Na_2CO_3 extract is added to about 5 ml. potassium dichromate solution and heated.	Reaction mixture becomes green.	SO_3^{2-} present
11. Test of carbonate in presence of sulphite:		
The solid salt mixture is taken in a test tube. Dilute hydrochloric acid is added to it. The gas evolved is passed through a boiling tube containing potassium dichromate solution and dilute H_2SO_4. The residual gas is then passed through lime water.	Lime water becomes milky.	CO_3^{2-} present
12. Silver nitrate test with neutral extract:		
About 5 c.c. sodium carbonate extract is taken in a test tube. Dilute HNO_3 is added to it slowly till faintly acidic. It is tested by blue litmus paper which becomes red. The solution is boiled to expel all CO_2 gas. It is cooled and dilute NH_4OH is added	(i) Curdy white ppt., soluble in dil. NH_4OH but insoluble in dil. HNO_3.	Cl^- and/or SCN^- present
	(ii) White ppt., insoluble in dil. HNO_3 and dil. NH_4OH.	$[Fe(CN)_6]^{4-}$ present
	(iii) White ppt., soluble in dil. HNO_3 and dil. NH_4OH.	COO^- | COO^- and/or BO_3^{3-} and/or

(*Contd...*)

Experiment	Observation	Inference
dropwise till faintly alkaline. It is tested by red litmus paper which becomes blue. The solution is boiled till neutral (filter if precipitate appears). Now silver nitrate solution is added to this neutral extract.		$CH(OH)COO^-$ | $CH(OH)COO^-$ present
	(iv) Pale yellow ppt., insoluble in dil. HNO_3 and also in dil. NH_4OH but soluble in liquor ammonia.	Br^- present
	(v) Yellow ppt., insoluble in dil. HNO_3 and even in liquor ammonia.	I^- present
	(vi) Yellow ppt., soluble in dil. HNO_3 and dil. NH_4OH.	PO_4^{3-} and/or ASO_3^{3-} present
	(vii) Black ppt., insoluble in dil. HNO_3 but soluble in hot and conc. HNO_3.	S^{2-} present
	(viii) Bright red (scarlet) ppt., soluble in dil. HNO_3 and NH_4OH.	CrO_4^{2-} present
	(ix) Reddish brown ppt., soluble in dil. HNO_3 and dil. NH_4OH.	AsO_4^{3-} present
	(x) Orange ppt., insoluble in dil. HNO_3 and soluble in dil NH_4OH.	$\left[Fe^{III}(CN)_6\right]^{3-}$ present
	(xi) White ppt., gradually changes to dark.	$S_2O_3^{2-}$ Present
13. Ferric chloride test with neutral extract:		
About 5 c.c. neutral sodium carbonate extract (prepared as above) is taken. Neutral ferric chloride solution is added	(i) Blue ppt.	$\left[Fe^{II}(CN)_6\right]^{4-}$ present
	(ii) Blood red colouration, discharged by $HgCl_2$ solution.	SCN^- present

(*Contd...*)

Experiment	Observation	Inference
dropwise till change occurs.	(iii) Reddish purple colouration, discharged on warming.	$S_2O_3^{2-}$ present
	(iv) Brown colouration, blue ppt. with $FeSO_4$ solution.	$[Fe^{III}(CN)_6]^{3-}$ present
14. Blue ring test:		
The neutral sodium carbonate extract (prepared as before) is mixed with ammonium molybdate solution. The resulting solution is poured slowly down the side of a test tube which contains conc. H_2SO_4.	Temporary blue ring at the contact zone.	$S_2O_3^{2-}$ present
15. Fenton's test:		
About 5 c.c. neutral extract is taken. One drop of saturated $FeSO_4$ solution is added to it. Now 2-3 drops of hydrogen peroxide (10 volume) is added. At last excess sodium hydroxide soln. is added.	Intense violet colouration.	CH(OH)COO⁻ | CH(OH)COO⁻ present
16. Ammonium chloride test with neutral extract:		
About 10 c.c. neutral extract is taken. It is evaporated to half its bulk. Now the extract is cooled and strong ammonium chloride solution is added to it.	White gelatinous precipitate.	SiO_3^{2-} present
17. Ammonium molybdate and stannous chloride test:		

(*Contd...*)

Experiment	Observation	Inference
About 5 c.c. neutral extract is taken. About 10 c.c. ammonium molybdate (free from HNO_3) solution is added to it. Now dil. HCl is added dropwise till faintly acidic. At last a few c.c. of stannous chloride is added.	Deep blue colouration.	SiO_3^{2-} present
18. Microcosmic salt bead test:		
A transparent bead of microcosmic salt is made on the loop of platinum wire. Few particles of solid salt mixture are taken on it and heated.	Produced silica swims about in the bead.	SiO_3^{2-} present
19. Ammonium molybdate test:		
About 5 c.c. sodium carbonate extract is taken in a test tube. The extract is acidified with conc. HNO_3. Finally 10 c.c. ammonium molybdate is added and (a) warmed to about 40°C	(a) Canary-yellow precipitate.	PO_4^{3-} present
(b) boiled.	(b) Canary-yellow ppt.	AsO_4^{3-} present
20. Test of phosphate in presence of arsenate:		
About 1 c.c. of 10% ammonium iodide and 4 c.c. conc. HCl are added to the neutral extract and heated to boiling. Now H_2S is passed for 10 minutes. It is filtered. The filtrate is heated to dryness. The remaining mass is dissolved in dil. HNO_3	Canary-yellow ppt.	PO_4^{3-} present

(Contd...)

Experiment	Observation	Inference
and ammonium molybdate solution is added. It is warmed to 40°C.		
21. Test of borate:		
Methyl alcohol (or ethyl alcohol) and conc. H_2SO_4 are mixed with solid salt mixture to make a paste. The paste is taken on a rod and burnt in a flame.	Green flame	BO_3^{3-} present
If copper and barium are present, following test is performed:		
A paste of inorganic mixture, calcium fluoride and a few drops of conc. H_2SO_4 is prepared. A little of this paste is brought upon the end of a glass rod very close to the base of a Bunsen flame without actually touching it.	Green flame	BO_3^{3-} present
22. Test of oxidising ion:		
(a) The solution of the substance (or sodium carbonate extract of the substance) is acidified with dil. HCl and H_2S gas passed.	(a) White or pale yellow colloidal ppt. of sulphur or change in colour of solution.	Oxidising agents.
(b) The substance is gently heated with a little conc. HCl.	(b) Evolution of greenish yellow gas (chlorine) having suffocating odour.	Oxidising agents (e.g., chromates, dichromates and permanganates)
(c) Colour of the Na_2CO_3 extract of the substance observed.	(c) (i) Violet/Pink	Permanganates (MnO_4^-)
	(ii) Orange yellow	Dichromates $(Cr_2O_7^{2-})$

(Contd...)

Experiment	Observation	Inference
	(iii) Yellow	Chromates (CrO_4^{2-})
(d) H_2S gas is passed through the sodium carbonate extract and the change in colour of the extract observed.	(d) (i) Violet → Colourless	MnO_4^- present
	(ii) Orange yellow → Dirty green	$Cr_2O_7^{2-}$ present
	(iii) Yellow → Dirty green	CrO_4^{2-} present
(e) The orange yellow/ yellow sodium extract is acidified with dilute acetic acid and lead acetate solution added.	(e) A yellow ppt. of $PbCrO_4$.	CrO_4^{2-} and $Cr_2O_7^{2-}$
23. Hydrogen peroxide test for vanadate:		
About 5 c.c. sodium carbonate extract is taken in a test tube. It is acidified with dil. H_2SO_4 and then a few drops of hydrogen peroxide are added (avoid excess).	Red colouration	VO_3^- present
24. Barium chloride test for vanadate:		
About 5 c.c. sodium carbonate extract is taken in a test tube and barium chloride solution is added to it.	Yellow ppt.; soluble in dil. HCl.	VO_3^- present
25. Test of tungstate:		
About 5 c.c. sodium carbonate extract is taken in a test tube. It is acidified with dil. HCl. Now 1-2 pieces of zinc metal are added.	Blue colouration (or precipitate)	WO_4^{2-} present

(*Contd...*)

Experiment	Observation	Inference
26. Test of tungstate and molybdate:		
About 5 c.c. sodium carbonate extract is taken in a test tube. Now tin (II) chloride and conc. HCl are added and heated.	(i) Blue colouration. (ii) Blue colouration which changes to green and finally brown.	WO_4^{2-} present MoO_4^{2-} present
27. Test of molybdate:		
About 5 c.c. sodium carbonate extract is taken in a test tube. The extract is acidified with dil. HCl. Now tin (II) chloride solution and ammonium thiocyanate (or potassium thiocyanate) solution are added.	Red colouration	MoO_4^{2-} present
28. Test for acetate:		
(i) A small amount of the mixture is treated with dil. H_2SO_4 and warmed gently.	Vinegar-like odour	CH_3COO^- (acetate) present
(ii) An aqueous solution of the mixture is prepared or if the mixture is soluble in dilute acid, the acidic solution is neutralised with ammonia and neutral $FeCl_3$ solution added.	Deep red colouration	CH_3COO^- (acetate) present
(iii) The above solution is diluted and heated to boiling.	Brown-red ppt.	CH_3COO^- (acetate) present
29. To a small portion of the Na_2CO_3 extract is added a few drops of a solution of sodium nitroprusside.	A violet or pink colour is developed.	S^{2-} present

5.3.1. Explanation of barium chloride test

About 5 c.c. sodium carbonate extract is taken in a test tube. Dilute hydrochloric acid is added slowly with stirring till the solution turns distinctly acidic (blue litmus turns red). It is boiled to expel all the CO_2. The solution is cooled. Excess of barium chloride solution is added to it. Barium ion of the reagent combines with sulphate ion of the extract to produce insoluble white barium sulphate.

$$Ba^{2+} + SO_4^{2-} \rightarrow \underset{\text{(White ppt.)}}{BaSO_4} \downarrow$$

Things to Remember

(i) Dilute HNO_3 may be used instead of dil. HCl.

(ii) Dilute H_2SO_4 can't be used because we are doing the test of SO_4^{2-}.

(iii) Conc. HCl or conc. HNO_3 should not be used, as a precipitate of barium chloride or of barium nitrate may form.

(iv) Under the condition of the experiment carbonate, sulphite and phosphate are not precipitated.

5.3.2. Explanation of silver nitrate test

About 5 c.c. sodium carbonate extract is taken in a test tube. Dilute nitric acid is added with stirring till the solution turns distinctly acidic (blue litmus turns red). It is boiled for 2-3 minutes to expel all the carbon dioxide. The solution is cooled. Now excess of silver nitrate solution is added to it. Consequently, precipitation of silver halide or silver sulphide takes place if halide or sulphide ion is present in the Na_2CO_3 extract.

$$Cl^- + Ag^+ \rightarrow \underset{\text{(White curdy ppt.)}}{AgCl}$$

$$Br^- + Ag^+ \rightarrow \underset{\text{(Pale yellow ppt.)}}{AgBr}$$

$$I^- + Ag^+ \rightarrow \underset{\text{(Yellow ppt.)}}{AgI}$$

$$S^{2-} + 2Ag^+ \rightarrow \underset{\text{(Brown or black ppt.)}}{Ag_2S}$$

Silver chloride dissolves in dil. NH_4OH forming a complex, diammine silver (I) chloride.

$$AgCl + 2NH_3 \rightarrow \left[Ag\left(NH_3\right)_2\right]Cl$$

Silver bromide is insoluble in dil. NH_4OH and soluble in liquor ammonia forming a complex, diammine silver (I) bromide.

$$AgBr + 2NH_3 \rightarrow \left[Ag\left(NH_3\right)_2\right]Br$$

Silver iodide neither dissolves in dil. NH_4OH nor in liquor ammonia.

$$AgI + NH_3 \rightarrow \text{No complex formation}$$

Things to Remember

(i) Treatment of ammonia solution in the above test is unnecessary when more than one halide is present. Thus such test is not utilized as confirmatory test.

(ii) Silver fluoride is soluble in water.

5.3.3. Explanation of chromyl chloride test

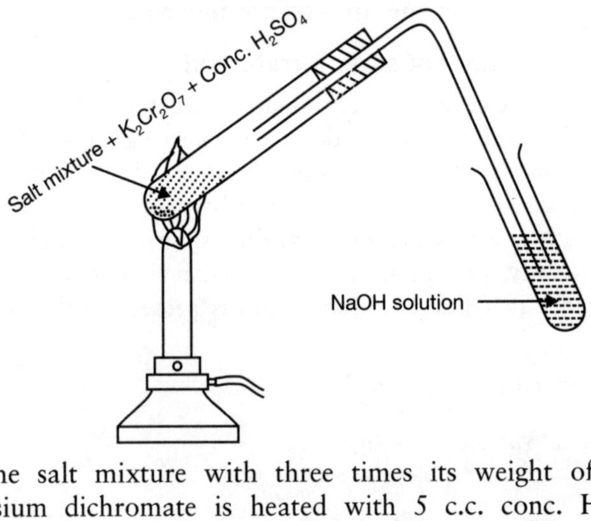

The salt mixture with three times its weight of solid potassium dichromate is heated with 5 c.c. conc. H_2SO_4. Consequently, red vapour of chromyl chloride evolves.

$$Cr_2O_7^{2-} + 4Cl^- + 6H^+ \rightarrow \underset{\text{(Red vapour)}}{2CrO_2Cl_2} \uparrow + 3H_2O$$

This red vapour of chromyl chloride is passed through sodium hydroxide solution, when a yellow solution of sodium chromate is obtained. The yellow colour is due to formation of chromate ion.

$$CrO_2Cl_2 + 4OH^- \rightarrow \underset{\text{(Yellow)}}{CrO_4^{2-}} + 2Cl^- + 2H_2O$$

Now the yellow solution is acidified with dilute acetic acid and then lead acetate solution is added. Consequently, yellow precipitate of lead chromate is obtained.

$$CrO_4^{2-} + Pb^{2+} \rightarrow \underset{\text{(Yellow ppt.)}}{PbCrO_4}$$

Things to Remember

(i) Bromide and iodide give off halogens which yield colourless solution with sodium hydroxide solution, and hence this test may be employed for identification of chloride in presence of bromide and iodide.

$$Cr_2O_7^{2-} + 14H^+ + 6\not{e} \rightarrow 2Cr^{3+} + 7H_2O$$

$$[2X^- \rightarrow X_2 + 2\not{e}] \times 3$$

$$\overline{Cr_2O_7^{2-} + 6X^- + 14H^+ \rightarrow 2Cr^{3+} + 7H_2O + 3X_2}$$

where X = Br or I

(ii) Fluoride gives volatile liquid chromyl fluoride, CrO_2F_2, which decomposes by water. Thus fluoride must be absent for this test. If fluoride be present, it should be removed before this test. Removal of fluoride has been given later.

5.3.4. Explanation of test of bromide

About 5 c.c. sodium carbonate extract is taken in a test tube and acidified with dil. HCl (test with blue litmus). The solution is boiled to expel all the carbon dioxide gas and cooled. It is shaken with few drops of carbon tetrachloride (or chloroform or carbon disulphide) and chlorine water. Organic layer becomes yellow or orange. It is due to formation of bromine.

$$2Br^- \rightarrow Br_2 + 2\cancel{e}$$

$$Cl_2 + 2\cancel{e} \rightarrow 2Cl^-$$

$$2Br^- + Cl_2 \rightarrow \underset{\text{(Yellow/brown)}}{Br_2} + 2Cl^-$$

Things to Remember

(i) All halogens are oxidising agent and oxidising power decreases from top to bottom in halogen group of P.T.

(ii) Upper halogen of the P.T. oxidises the lower halide of the same.

5.3.5. Explanation of test of iodide

About 5 c.c. sodium carbonate extract is taken in a test tube and acidified with dil. HCl (test with blue litmus). The solution is boiled to expel all carbon dioxide gas and cooled. It is shaken with few drops of carbon tetrachloride (or chloroform or carbon disulphide) and chlorine water. Violet (purple) colour appears in organic layer. It is due to formation of iodine.

$$2I^- \rightarrow I_2 + 2\cancel{e}$$

$$Cl_2 + 2\cancel{e} \rightarrow 2Cl^-$$

$$2I^- + Cl_2 \rightarrow \underset{\text{(Violet)}}{I_2} + 2Cl^-$$

5.3.6. Explanation of test of bromide and iodide in presence of each other

About 5 c.c. sodium carbonate extract is taken in a test tube and acidified with dil. HCl. The solution is boiled to expel all the carbon dioxide and cooled. Now 1-2 c.c. chloroform (or carbon tetrachloride) is added and then shaken after the addition of 1-2 c.c. chlorine water. Organic layer becomes violet (purple) due to production of iodine.

$$2I^- \rightarrow I_2 + 2\cancel{e}$$

$$Cl_2 + 2\cancel{e} \rightarrow 2Cl^-$$

$$2I^- + Cl_2 \rightarrow \underset{\text{(Violet)}}{I_2} + 2Cl^-$$

Now chlorine water is added drop by drop till the disappearance of violet colour. This is due to oxidation of iodine into iodate.

$$I_2 + 6H_2O \rightarrow 2IO_3^- + 12H^+ + 10\cancel{e}$$
$$\left[Cl_2 + 2\cancel{e} \rightarrow 2Cl^-\right] \times 5$$
$$\overline{I_2 + 6H_2O + 5Cl_2 \rightarrow \underset{\text{(Colourless)}}{2IO_3^-} + 12H^+ + 10Cl^-}$$

Further addition of chlorine water produces brown colouration in organic layer if bromide is present, otherwise the solution remains colourless.

$$2Br^- \rightarrow Br_2 + 2\cancel{e}$$
$$Cl_2 + 2\cancel{e} \rightarrow 2Cl^-$$
$$\overline{2Br^- + Cl_2 \rightarrow \underset{\text{(Brown)}}{Br_2} + 2Cl^-}$$

Things to Remember

(i) This test is applicable even in presence of chloride.

(ii) In the above test chlorine oxidises iodide first because iodide has more reducing power than bromide.

5.3.7. Explanation of ring test

A pinch of salt mixture is taken in a test tube. About 5 c.c. water is added in it. It is shaken. If the mixture dissolves completely, the solution is used to perform the ring test. If the mixture does not dissolve completely, it is filtered and the filtrate is used to perform the ring test. It should be noted that almost all nitrates are soluble in water. Alternatively, Na_2CO_3 extract of the salt mixture is first acidified with dil. H_2SO_4 and used for the test.

For preparation of fresh $FeSO_4$ solution, few crystals of $FeSO_4$ (light green) are taken in a test tube. 10 c.c. water is added in it. It is shaken and decanted. This process is repeated until we have light green crystals of $FeSO_4$. It is done so because $FeSO_4$ kept in laboratory acquires a layer of ferric sulphate (brown) due to aerial oxidation.

$$\underset{\text{(Light green)}}{Fe^{2+}} \rightarrow \underset{\text{(Brown)}}{Fe^{3+}} + e$$

Thus to remove ferric sulphate, it is essential to wash the crystals of ferrous sulphate repeatedly.

Now 5 c.c. water is added to green crystals of ferrous sulphate so formed and shaken to dissolve (never heat). This is called freshly prepared ferrous sulphate solution.

Ferrous sulphate solution thus prepared must be sufficiently concentrated, preferably saturated. Now this freshly prepared strong solution of ferrous sulphate is added into the cold aqueous solution of salt mixture or its acidified Na_2CO_3 extract. The test tube containing the mixture of ferrous sulphate and the salt solution is held firmly in a slanting position, then concentrated sulphuric acid is added very cautiously along the side of the test tube so that the acid being heavy, settles down to the bottom and forms a separate layer underneath the solution about half an inch deep. The test tube should not be shaken, otherwise the acid and the solution will be mixed and thus heat will be generated and the test will be spoiled. A brown ring will be seen at the junction of the acid and the solution proving the presence of nitrate. If nitrite is present, the whole solution turns brown.

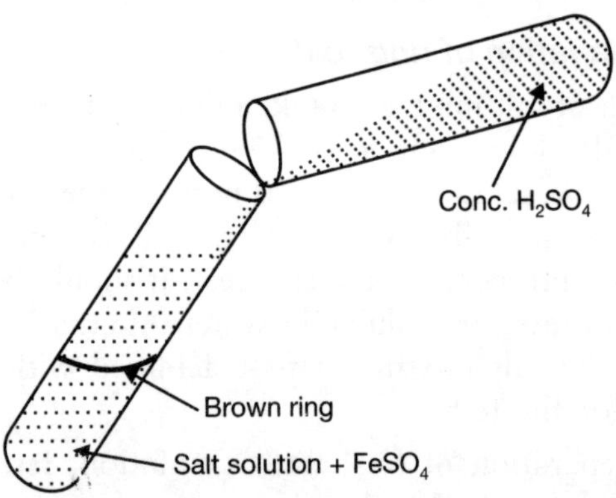

The brown ring is due to the formation of the $[Fe^{I}(NO)]^{2+}$. This can be explained as under:

Sulphuric acid liberates nitric acid from nitrate.

$NO_3^- + H^+ \rightarrow HNO_3$

Nitric acid oxidises ferrous sulphate into ferric sulphate, and is reduced itself into nitric oxide.

$$NO_3^- + 4H^+ + 3\cancel{e} \rightarrow NO + 2H_2O$$

$$\left[Fe^{2+} \rightarrow Fe^{3+} + \cancel{e}\right] \times 3$$

$$3Fe^{2+} + NO_3^- + 4H^+ \rightarrow 3Fe^{3+} + 2H_2O + NO$$

Nitric oxide unites with the still unoxidised (excess) portion of the ferrous sulphate.

$$Fe^{2+} + NO \rightarrow \left[Fe^{I}(NO)\right]^{2+}$$

The univalent character of iron in the complex is justified by the presence of a co-ordinated NO^+ group.

$$\left[Fe^{2+}(NO)\right]^{2+} \rightarrow \left[Fe^{+}(NO^{+})\right]^{2+}$$

Note: Solution of nitrites acidified even with dil. H_2SO_4 when treated with ferrous sulphate solution form the brown complex, $[Fe(NO)]SO_4$.

$$NaNO_2 + H_2SO_4 \rightarrow HNO_2 + NaHSO_4$$

$$3HNO_2 \rightarrow H_2O + HNO_3 + 2NO$$

$$FeSO_4 + NO \rightarrow [Fe(NO)]\,SO_4$$

Things to Remember

(i) All nitrates are soluble in water except Bi^{3+}, Hg^+ and Hg^{2+} which give white ppt. of basic nitrate when treated with excess of water. In such cases, salt should be treated with water, shaken and filtered. And with this aqueous extract ring test should be performed.

(ii) Ferrous sulphate solution should be free from ferric sulphate. It should be freshly prepared and must be fairly concentrated.

(iii) If a ppt. is formed on addition of ferrous sulphate solution to the nitrate solution, we should filter it and proceed with the filtrate.

(iv) This test is done only when nitrate is suspected from conc. sulphuric acid test.

(v) This test is unreliable in presence of bromide, iodide, nitrite, chromate, etc. Bromide and iodide interfere because of the liberated halogen.

5.3.8. *Explanation of test of nitrate in presence of bromide and iodide*

The sodium carbonate extract is boiled with sodium hydroxide solution until ammonium salts, if present, are completely decomposed.

$$NH_4^+ + OH^- \rightarrow NH_3 \uparrow + H_2O$$

Now the solution is cooled under the tap. Powdered Devarda's alloy or aluminium powder or zinc dust is then added and the mixture gently warmed. The evolution of ammonia, detected by its smell and its action upon red litmus paper indicates the presence of a nitrate.

$$\left[Zn + 4OH^- \rightarrow ZnO_2^{2-} + 2H_2O + 2e\!\!\!/\right] \times 4$$

$$NO_3^- + 3OH^- + 6H_2O + 8e\!\!\!/ \rightarrow NH_3 + 12OH^-$$

$$\overline{4Zn + NO_3^- + 7OH^- \rightarrow 4ZnO_2^{2-} + 2H_2O + NH_3 \uparrow}$$

Things to Remember

(i) The brown ring test for nitrate can't be applied in presence of bromide and iodide since the liberation of free halogen with concentrated sulphuric acid (O.A.) will obscure the brown ring due to nitrate.

$$SO_4^{2-} + 4H^+ + 2e\!\!\!/ \rightarrow SO_2 + 2H_2O$$

$$2X^- \rightarrow X_2 + 2e\!\!\!/$$

$$\overline{SO_4^{2-} + 2X^- + 4H^+ \rightarrow 2H_2O + SO_2 + X_2}$$

where, X = Br, I

(ii) Devarda's alloy contains 50% copper, 45% aluminium and 5% zinc.

(iii) Devarda's alloy is a reducing agent which reduces nitrate into ammonia.

5.3.9. *Explanation of test of nitrate in presence of nitrite*

To decompose nitrite, solid salt mixture and solid ammonium chloride in the ratio of 1:3 are taken. A little water is added and boiled till effervescence ceases. The effervescence is due to evolution of nitrogen by decomposition of nitrite.

$$NH_4^+ + NO_2^- \rightarrow NH_4\cancel{NO}_2$$
$$NH_4\cancel{NO}_2 \rightarrow 2H_2O + N_2$$
$$\overline{NH_4^+ + NO_2^- \rightarrow 2H_2O + N_2 \uparrow}$$

The residue is treated with water and filtered. Now ring test for nitrate is done with this filtrate.

Things to Remember

(i) The nitrite is readily identified in the presence of a nitrate by treatment with dilute HCl or H_2SO_4.

(ii) The nitrate can't be detected in presence of nitrite since the latter also gives the brown ring test. Thus, before ring test for nitrate, it is necessary to eliminate the presence of nitrite.

5.3.10. Explanation of test of sulphite

About 5 c.c. sodium carbonate extract is added to about 5 c.c. potassium dichromate solution and warmed. The reaction mixture becomes green due to reduction of $Cr_2O_7^{2-}$ into Cr^{3+} by SO_3^{2-} (sulphite ion)

$$Cr_2O_7^{2-} + 14H^+ + 6\cancel{e} \rightarrow 2Cr^{3+} + 7H_2O$$
$$[SO_3^{2-} + H_2O \rightarrow SO_4^{2-} + 2H^+ + 2\cancel{e}] \times 3$$
$$\overline{Cr_2O_7^{2-} + 3SO_3^{2-} + 8H^+ \rightarrow \underset{\text{(Green)}}{2Cr^{3+}} + 3SO_4^{2-} + 4H_2O}$$

Things to Remember

(i) This test is only applicable when halides or other reducing agents are absent.

(ii) In case of presence of any reducing agent, presence of sulphite should be judged in dilute hydrochloric acid test.

(iii) Presence of carbonate does not interfere this test.

5.3.11. Explanation of test of carbonate in presence of sulphite

The solid salt mixture is taken in a test tube and dilute hydrochloric acid is added. Consequently, SO_2 and CO_2 gases are formed.

$$SO_3^{2-} + 2H^+ \rightarrow H_2O + SO_2 \uparrow$$

$$CO_3^{2-} + 2H^+ \rightarrow H_2O + CO_2 \uparrow$$

The gases are allowed to pass first through a boiling tube containing $K_2Cr_2O_7$ solution and sulphuric acid. Sulphur dioxide is absorbed here and the solution of boiling tube becomes green. The remaining gas, carbon dioxide, is passed through lime water or baryta water. The carbon dioxide turns lime water or baryta water milky.

$$Ca(OH)_2 + CO_2 \rightarrow \underset{\text{(White ppt.)}}{CaCO_3} + H_2O$$

$$Ba(OH)_2 + CO_2 \rightarrow \underset{\text{(White ppt.)}}{BaCO_3} + H_2O$$

Things to Remember

(i) An alternative procedure is to add a little powdered potassium dichromate to the salt mixture and then to warm with dilute sulphuric acid; the evolved gas is passed through lime water. This method requires shorter time.

(ii) The above method can, of course, be applied in presence of thiosulphate.

$$S_2O_3^{2-} + 2H^+ \rightarrow S \downarrow + SO_2 \uparrow + H_2O$$

(iii) Saturated aqueous solution of barium hydroxide is known as baryta solution.

5.3.12. Explanation of silver nitrate test with neutral extract

About 5 c.c. sodium carbonate extract is taken in a basin. Dilute nitric acid is added slowly till faintly acidic (test by blue litmus paper). The solution is boiled to expel all carbon dioxide. It is cooled and dilute ammonium hydroxide is added drop by drop to make it faintly alkaline (test by red litmus paper). The solution is boiled to expel excess ammonia to make it neutral. Now silver nitrate solution is added to this neutral extract.

(i) White curdy precipitate indicates the presence of chloride and thiocyanate.

$$Cl^- + Ag^+ \rightarrow \underset{\text{(White curdy ppt.)}}{AgCl}$$

$$SCN^- + Ag^+ \rightarrow \underset{\text{(White curdy ppt.)}}{AgSCN}$$

The precipitate is soluble in dilute ammonium hydroxide due to formation of a complex.

$$AgCl + 2NH_3 \rightarrow \underset{\text{[Diammine silver (I) chloride]}}{[Ag(NH_3)_2]^+Cl^-}$$

$$AgSCN + 2NH_3 \rightarrow [Ag(NH_3)_2]^+SCN^-$$

Silver chloride and silver thiocyanate are insoluble in dilute nitric acid.

(ii) White precipitate being insoluble in dilute HNO_3 and dilute ammonium hydroxide indicates the presence of ferrocyanide.

$$[Fe^{II}(CN)_6]^{4-} + 4Ag^+ \rightarrow \underset{\text{[Silver hexacyano ferrate (II)]}}{Ag_4[Fe^{II}(CN)_6]\downarrow}$$

(iii) White precipitate being soluble in dilute HNO_3 and dil. NH_4OH indicates the presence of oxalate, tartarate and borate.

$$\underset{\text{(Oxalate)}}{\begin{matrix} COO^- \\ | \\ COO^- \end{matrix}} + 2Ag^+ \rightarrow \underset{\text{(White ppt.)}}{\begin{matrix} COOAg \\ | \\ COOAg \end{matrix}}$$

$$\begin{matrix} COOAg \\ | \\ COOAg \end{matrix} + 4NH_3 \rightarrow 2\underset{\text{(Soluble)}}{[Ag(NH_3)_2]^+} + \underset{\text{(Soluble)}}{\begin{matrix} COO^- \\ | \\ COO^- \end{matrix}}$$

$$\begin{matrix} COOAg \\ | \\ COOAg \end{matrix} + 2H^+ \rightarrow \underset{\text{(Soluble)}}{2Ag^+} + \underset{\text{(Soluble)}}{\begin{matrix} COOH \\ | \\ COOH \end{matrix}}$$

We can represent the reactions for tartarate as follows:

$$\underset{\text{(Tartarate)}}{\begin{matrix} CH(OH)COO^- \\ | \\ CH(OH)COO^- \end{matrix}} + 2Ag^+ \rightarrow \underset{\text{(White ppt.)}}{\begin{matrix} CH(OH)COOAg \\ | \\ CH(OH)COOAg \end{matrix}}$$

$$\begin{array}{l} CH(OH)COOAg \\ | \\ CH(OH)COOAg \end{array} + 4NH_3 \rightarrow \underset{\text{(Soluble)}}{2[Ag(NH_3)_2]^+} + \underset{\text{(Soluble)}}{\begin{array}{l} CH(OH)COO^- \\ | \\ CH(OH)COO^- \end{array}}$$

$$\begin{array}{l} CH(OH)\,COOAg \\ | \\ CH(OH)\,COOAg \end{array} + 2H^+ \rightarrow \underset{\text{(Soluble)}}{2Ag^+} + \underset{\text{(Soluble)}}{\begin{array}{l} CH(OH)COOH \\ | \\ CH(OH)COOH \end{array}}$$

In a similar fashion, reactions of borate can be depicted as:

$$\underset{\text{(Borate)}}{BO_3^{3-}} + Ag^+ + H_2O \rightarrow \underset{\substack{\text{Silver metaborate} \\ \text{(White ppt.)}}}{AgBO_2} + 2OH^-$$

$$AgBO_2 + 2NH_3 + 2OH^- \rightarrow \underset{\text{(Soluble)}}{[Ag(NH_3)_2]^+} + \underset{\text{(Soluble)}}{BO_3^{3-}} + H_2O$$

$$AgBO_2 + H_2O \rightarrow \underset{\text{(Soluble)}}{Ag^+} + \underset{\text{(Soluble)}}{BO_3^{3-}} + 2H^+$$

(iv) Pale yellow precipitate of silver bromide being soluble in liquor ammonia indicates the presence of bromide.

$$Br^- + Ag^+ \rightarrow \underset{\text{(Pale yellow ppt.)}}{AgBr}$$

$$AgBr + 2NH_3 \rightarrow \underset{\text{(Soluble)}}{[Ag(NH_3)_2]}Br$$

(v) Yellow precipitate of silver iodide being insoluble in dilute ammonium hydroxide or liquor ammonia, indicates the presence of iodide.

$$I^- + Ag^+ \rightarrow AgI$$

$$AgI + NH_3 \rightarrow \text{No complex formation}$$

(vi) Yellow precipitate being soluble in dil. HNO_3 and dilute NH_4OH indicates the presence of phosphate and arsenite.

$$PO_4^{3-} + 3Ag^+ \rightarrow \underset{\substack{\text{Silver phosphate} \\ \text{(Yellow precipitate)}}}{Ag_3PO_4\downarrow}$$

$$Ag_3PO_4 + 3H^+ \rightarrow \underset{\text{(Soluble)}}{3Ag^+} + \underset{\text{(Soluble)}}{H_3PO_4}$$

$$Ag_3PO_4 + 6NH_3 \rightarrow \underset{\text{(Soluble)}}{3[Ag(NH_3)_2]^+} + \underset{\text{(Soluble)}}{PO_4^{3-}}$$

In the similar fashion, reactions of arsenite can be given.

$$AsO_3^{3-} + 3Ag^+ \rightarrow \underset{\substack{\text{Silver arsenite} \\ \text{(Yellow ppt.)}}}{Ag_3AsO_3} \downarrow$$

$$Ag_3AsO_3 + 3H^+ \rightarrow \underset{\text{(Soluble)}}{3Ag^+} + \underset{\text{(Soluble)}}{H_3AsO_3}$$

$$Ag_3AsO_3 + 6NH_3 \rightarrow \underset{\text{(Soluble)}}{3\left[Ag(NH_3)_2\right]^+} + \underset{\text{(Soluble)}}{AsO_3^{3-}}$$

(vii) Black precipitate of silver sulphide being insoluble in dil. HNO_3 but soluble in hot and conc. HNO_3 indicates the presence of sulphide.

$$S^{2-} + 2Ag^+ \rightarrow Ag_2S$$

(viii) Bright red precipitate (scarlet) of silver chromate being soluble in dil. HNO_3 and NH_4OH indicates the presence of chromate.

$$CrO_4^{2-} + 2Ag^+ \rightarrow \underset{\substack{\text{Silver chromate} \\ \text{(Scarlet ppt.)}}}{Ag_2CrO_4} \downarrow$$

$$2Ag_2CrO_4 + 2H^+ \rightarrow \underset{\text{(Soluble)}}{4Ag^+} + \underset{\text{(Soluble)}}{Cr_2O_7^{2-}} + H_2O$$

$$Ag_2CrO_4 + 4NH_3 \rightarrow \underset{\text{(Soluble)}}{2\left[Ag(NH_3)_2\right]^+} + \underset{\text{(Soluble)}}{CrO_4^{2-}}$$

(ix) Reddish brown precipitate of silver arsenate being soluble in dil. HNO_3 and NH_4OH indicates the presence of arsenate.

$$AsO_4^{3-} + 3Ag^+ \rightarrow \underset{\substack{\text{Silver arsenate} \\ \text{(Brownish red ppt.)}}}{Ag_3AsO_4} \downarrow$$

$$Ag_3AsO_4 + 3H^+ \rightarrow \underset{\text{(Soluble)}}{3Ag^+} + \underset{\text{(Soluble)}}{H_3AsO_4}$$

$$Ag_3AsO_4 + 6NH_3 \rightarrow \underset{\text{(Soluble)}}{3\left[Ag(NH_3)_2\right]^+} + \underset{\text{(Soluble)}}{AsO_4^{3-}}$$

(x) Orange precipitate of silver ferricyanide being soluble in ammonia solution indicates the presence of ferricyanide ion.

$$\left[Fe^{III}(CN)_6\right]^{3-} + 3Ag^+ \rightarrow \underset{\substack{\text{Silver hexacyano ferrate (III)} \\ \text{(Orange ppt.)}}}{Ag_3\left[Fe^{III}(CN)_6\right]} \downarrow$$

$$Ag_3[Fe(CN)_6] + 6NH_3 \rightarrow 3\underset{\text{(Soluble)}}{[Ag(NH_3)_2]^+} + \underset{\text{(Soluble)}}{[Fe^{III}(CN)_6]^{3-}}$$

(xi) White precipitate of silver thiosulphate is formed.

$$S_2O_3^{2-} + 2Ag^+ \rightarrow \underset{\text{(White ppt.)}}{Ag_2S_2O_3} \downarrow$$

The precipitate is unstable, turning dark on standing, when silver sulphide is formed.

$$Ag_2S_2O_3 + H_2O \rightarrow Ag_2S \downarrow + H_2SO_4$$

This hydrolytic decomposition can be accelerated by warming.

Things to Remember

(i) Silver ferrocyanide is insoluble in dilute NH_4OH while silver ferricyanide is soluble. This, apart from the colour of the precipitate is distinction between the two.

(ii) Silver arsenite is yellow whereas silver arsenate is brownish red. This is distinction between arsenite and arsenate.

5.3.13. Explanation of ferric chloride test with neutral extract

Neutral ferric chloride solution is added to about 5 c.c. neutral 'extract' dropwise till the change occurs.

(i) Prussian blue precipitate confirms the presence of ferrocyanide.

$$3[Fe^{II}(CN)_6]^{4-} + 4Fe^{3+} \rightarrow \underset{\substack{\text{Ferriferrocyanide} \\ \text{(Prussian blue)}}}{Fe_4^{III}[Fe^{II}(CN)_6]_3} \downarrow$$

(ii) Blood red colouration due to formation of a complex confirms the presence of thiocyanate.

$$3SCN^- + Fe^{3+} \rightarrow [Fe(SCN)_3]$$

In fact there are a series of complex cations and anions formed. These complexes are:

$$[Fe^{III}(SCN)]^{2+}, [Fe^{III}(SCN)_2]^+, [Fe^{III}(SCN)_4]^-,$$

$$[Fe^{III}(SCN)_5]^{2-} \text{ and } [Fe^{III}(SCN)_6]^{3-}$$

(iii) Red purple colouration due to formation of dithiosulphate iron (III) complex confirms the presence of thiosulphate.

$$2S_2O_3^{2-} + Fe^{3+} \rightarrow \left[Fe^{III}\left(S_2O_3\right)_2\right]^-$$

On warming or even on standing the colour disappears rapidly, while tetrathionate and iron (II) ions are formed.

$$\left[Fe^{III}\left(S_2O_3\right)_2\right]^- + Fe^{3+} \rightarrow 2Fe^{2+} + S_4O_6^{2-}$$

(iv) Brown colouration, owing to the formation of undissociated iron (III) hexacyanoferrate (III).

$$\left[Fe^{III}(CN)_6\right]^{3-} + Fe^{3+} \rightarrow Fe^{III}\left[Fe^{III}(CN)_6\right]$$

5.3.14. Explanation of blue ring test

When neutral sodium carbonate extract mixed with ammonium molybdate solution is poured slowly down the side of a test tube which contains concentrated sulphuric acid, a blue ring is formed temporarily at the contact zone. This confirms the presence of thiosulphate. The origin of this colour is as yet unexplained.

5.3.15. Explanation of Fenton's test

About 1-2 drops saturated ferrous sulphate solution are added to the neutral sodium carbonate extract. Then 2-3 drops of hydrogen peroxide (10 volume) are added to it. At last, excess of sodium hydroxide solution is added. Consequently, intense violet colouration develops. This confirms the presence of tartarate.

Things to Remember

Mixtue of ferrous sulphate solution and hydrogen peroxide is known as Fenton's reagent. Actually it is an oxidising agent.

5.3.16. Explanation of ammonium chloride test with neutral extract

About 10 c.c. neutral sodium carbonate extract is evaporated to make its volume half. Now the extract is cooled and strong ammonium chloride solution is added to it. White gelatinous precipitate is obtained which indicates the presence of silicate.

$$SiO_3^{2-} + 2NH_4^+ \rightarrow \underset{\text{(Gelatinous ppt. of silicic acid)}}{H_2SiO_3 \downarrow} + 2NH_3$$

5.3.17. Explanation of ammonium molybdate and stannous chloride test

About 10 c.c. ammonium molybdate (free from HNO_3) solution is added to about 5 c.c. neutral sodium carbonate extract. Now dil. HCl is added dropwise till faintly acidic. At last a few c.c. of stannous chloride is added. Production of deep blue colouration indicates the presence of silicate.

Ammonium molybdate reacts with silicate to form ammonium salt of silicomolybdic acid, $H_4[SiMo_{12}O_{40}]$. It is reduced by stannous chloride to produce molybdenum blue.

5.3.18. Explanation of microcosmic salt bead test

The microcosmic salt first fuses to a transparent bead consisting largely of sodium metaphosphate.

$$\underset{\text{Microcosmic salt}}{Na(NH_4)HPO_4} \xrightarrow{\Delta} NaPO_3 + H_2O \uparrow + NH_3$$

When a minute quantity of the solid silicate is taken on the hot bead and the whole again heated, the silica produced will not dissolve in the bead, but will swim about in fused mass, and is visible as white opaque masses or 'skeletons' in both the fused and the cold bead.

$$CaSiO_3 + NaPO_3 \rightarrow CaNaPO_4 + SiO_2$$

5.3.19. Explanation of ammonium molybdate test

When sodium carbonate extract acidified with conc. HNO_3 is heated with ammonium molybdate solution, canary-yellow precipitate is formed. This confirms the presence of phosphate.

$$Na_2HPO_4 + 12(NH_4)_2MoO_4 + 23HNO_3 \rightarrow$$

$$\underset{\text{(Ammonium phosphomolybdate; canary yellow ppt.)}}{(NH_4)_3(PMo_{12}O_{40})} + 21NH_4NO_3 + 2NaNO_3 + 12H_2O$$

5.3.20. Explanation of test of phosphate in presence of arsenate

Since both phosphate and arsenate give yellow precipitate with ammonium molybdate solution and nitric acid, of course, the former on warming (40°C) and later on boiling, it is advisable to remove arsenate before testing for phosphate if both are present together.

To remove arsenate, about 1 c.c. of 10% ammonium iodide and 4 c.c. conc. HCl are added to the neutral sodium carbonate extract and heated to boiling. Now H_2S is passed for 10 minutes until precipitation is complete. Arsenious sulphide is filtered off and the filtrate is evaporated to dryness to remove chlorides. Now the residue is dissolved in dilute nitric acid and then ammonium molybdate solution is added to test phosphate.

$$2H_3AsO_4 \rightarrow As_2O_5 + 3H_2O$$

$$As_2O_5 + 4HI \rightarrow As_2O_3 + 2H_2O + 2I_2$$

$$As_2O_3 + 3H_2S \rightarrow As_2S_3 + 3H_2O$$

$$2I_2 + 2H_2S \rightarrow 4HI + 2S$$

5.3.21. Explanation of test of borate

Methyl alcohol (or ethyl alcohol) and conc. H_2SO_4 are mixed with solid salt mixture to make a paste. The paste is taken on a rod and burnt in a flame. Green flame due to volatile methyl or ethyl borate indicates the presence of borate.

$$Na_2B_4O_7 + H_2SO_4 + 5H_2O \rightarrow 4H_3BO_3 + Na_2SO_4$$

$$H_3BO_3 + 3CH_3OH \rightarrow B(OCH)_3 + 3H_2O$$

Things to Remember

Barium and copper can interfere this test as these may give a similar green flame. In the presence of these salts, the borate is mixed with powdered calcium fluoride and conc. H_2SO_4. The paste is brought near the Bunsen flame when the highly volatile BF_3 is formed which colours the flame green.

$$CaF_2 + H_2SO_4 \rightarrow CaSO_4 + 2HF$$

$$Na_2B_4O_7 + H_2SO_4 \rightarrow 2B_2O_3 + Na_2SO_4 + H_2O$$

$$B_2O_3 + 6HF \rightarrow 2BF_3 + 3H_2O$$

5.3.22. Explanation of test of oxidising ion

(a) When H_2S gas is bubbled through an acidified solution of an oxidising agent, colloidal sulphur is precipitated due to the oxidation of H_2S or a change in colour of the solution occurs owing to the reduction of the oxidising agent.

(b) Chromates, dichromates and permanganates oxidise HCl resulting in the evolution of greenish yellow chlorine gas.

(c) Sodium permanganates, dichromates and chromates (which are formed in the sodium carbonate extract) have the colour as indicated in the table.

(d) Hydrogen sulphide gas is passed through the sodium carbonate extract and the change in colour of the extract observed.

(i) Change from violet to colourless confirms the presence of permanganate.

$$\left[MnO_4^- + 8H^+ + 5e^- \rightarrow Mn^{2+} + 4H_2O\right] \times 2$$

$$\left[S^{2-} \rightarrow S + 2e^-\right] \times 5$$

$$\underset{\text{(Violet)}}{2MnO_4^-} + 5S^{2-} + 16H^+ \rightarrow \underset{\text{(colourless)}}{2Mn^{2+}} + 5S + 8H_2O$$

(ii) Change from orange yellow to dirty green confirms the presence of dichromate.

$$Cr_2O_7^{2-} + 14H^+ + 6e^- \rightarrow 2Cr^{3+} + 7H_2O$$

$$\left[S^{2-} \rightarrow S + 2e^-\right] \times 3$$

$$\underset{\text{(Orange yellow)}}{Cr_2O_7^{2-}} + 3S^{2-} + 14H^+ \rightarrow \underset{\text{(Green)}}{2Cr^{3+}} + 3S + 7H_2O$$

Dirty green is due to presence of sulphur with Cr^{3+} ion.

(iii) Change from yellow to dirty green confirms the presence of chromate

$$\left[CrO_4^{2-} + 8H^+ + 3e^- \rightarrow Cr^{3+} + 4H_2O\right] \times 2$$

$$\left[S^{2-} \rightarrow S + 2e^-\right] \times 3$$

$$2CrO_4^{2-} + 3S^{2-} + 16H^+ \rightarrow \underset{\text{(Green)}}{2Cr^{3+}} + 3S + 8H_2O$$

(e) An orange yellow/yellow Na_2CO_3 extract indicates dichromates/chromates. The solution on acidification with dil.

acetic acid and treatment with lead acetate solution gives a yellow precipitate of $PbCrO_4$.

5.3.23. Explanation of hydrogen peroxide test for vanadate

About 5 c.c. sodium carbonate extract is acidified with dil. H_2SO_4 and then a few drops of hydrogen peroxide are added (avoid excess). Production of red colour confirms the presence of vanadate. The red colour is due to the formation of the peroxovanadium (V) cation, VO_2^{3+}.

$$\underset{\text{(Vanadate)}}{VO_3^-} + 4H^+ + H_2O_2 \rightarrow \underset{\text{(red)}}{VO_2^{3+}} + 3H_2O$$

If more hydrogen peroxide is added or the solution is made alkaline or both, the colour changes to yellow. The yellow colour is due to formation of diperoxoorthovanadate (V) ion, $\left[VO_2(O_2)_2\right]^{3-}$.

$$VO_2^{3+} + H_2O_2 + 6OH^- \rightarrow \left[VO_2(O_2)_2\right]^{3-} + 4H_2O$$

5.3.24. Explanation of barium chloride test for vanadate

Barium chloride solution is added to about 5 c.c. sodium carbonate extract. Yellow precipitate being soluble in dil. HCl indicates the presence of vanadate.

$$2VO_3^- + Ba^{2+} \rightarrow \underset{\text{(Yellow ppt.)}}{Ba(VO_3)_2} \downarrow$$

$$Ba(VO_3)_2 + 2H^+ \rightarrow \underset{\text{(Soluble)}}{Ba^{2+}} + \underset{\text{(Soluble)}}{2VO_3^-} + 2H^+$$

5.3.25. Explanation of test of tungstate

About 5 c.c. sodium carbonate is acidified with dil. HCl. Now 1-2 pieces zinc metal are added. Blue colouration or precipitate confirms the presence of tungstate. The product is called 'tungsten blue' and has a composition near to the formula W_2O_5.

5.3.26. Explanation of test of tungstate and molybdate

About 5 c.c. sodium carbonate extract is heated with stannous chloride and conc. HCl.

(i) Blue colouration indicates tungstate.

(ii) Blue colouration which changes to green and finally brown indicates molybdate.

5.3.27. Explanation of test of molybdate

About 5 c.c. sodium carbonate extract is acidified with dil. HCl. Then stannous chloride and ammonium thiocyanate (or potassium thiocyanate) solution are added. Red colouration indicates molybdate. Red colour is due to formation of hexathiocyanato molybdate (III), $[Mo(SCN)_6]^{3-}$.

5.3.28. Explanation of test of acetate

(i) Dilute sulfuric acid forms acetic acid when treated with acetates. Acetic acid is easily recognised by its vinegar like smell when the reaction solution is warmed.

$$CH_3COONa + H_2SO_4 \rightarrow NaHSO_4 + CH_3COOH$$

(ii) Neutral solutions of acetates on treatment with ferric chloride result in a deep red colouration due to the formation of ferric acetate.

$$FeCl_3 + 3\ CH_3{-}C\begin{matrix}\nearrow O \\ \searrow ONa\end{matrix} \rightarrow Fe(OCOCH_3)_3 + 3NaCl$$

(iii) On diluting and boiling the red solution, brown red ppt. of basic ferric acetate is formed.

$$Fe(OCOCH_3)_3 + 2H_2O \xrightarrow{\Delta} Fe(OH)_2 \cdot CH_3COO + 2CH_3COOH$$

5.3.29. Explanation of test of sulphide

A violet or pink colour is developed when the 'Na_2CO_3 extract' containing sulphide (Na_2S) is treated with a solution of sodium nitroprusside due to the formation of $Na_4[Fe(CN)_5NOS]$.

$$Na_2S + Na_2[Fe(CN)_5NO] \rightarrow Na_4[Fe(CN)_5NOS]$$

6. DRY TESTS FOR BASIC RADICALS

Dry tests are those tests which are directly applied to the given salt mixture.

To perform the dry tests for basic radicals, following tests can be done:

1. Action of heat

2. Charcoal reducing test
3. Charcoal oxidising test
4. Flame test
5. Borax bead test (for coloured salts only)

These tests have been comprised in Table 3 given below. Students are advised to follow the table only inside the laboratory and go through explanation of the tests, mentioned after the table, at home.

Table 3

Experiment	Observation	Inference
1. Action of heat:	*Colour changed*	
A pinch of the salt mixture is heated in a dry test tube.	Yellow (infusible) when hot, white when cold.	May be Zn^{2+}
	Yellow (fusible) when hot, white when cold.	May be Sb^{3+}
	White to yellowish brown (infusible) when hot, white or light yellow when cold.	May be Sn^{2+}, Sn^{4+}
	Yellow to reddish brown (infusible) when hot, yellow when cold.	May be Pb^{2+}
	Orange (fusible) when hot, yellow when cold.	May be Bi^{3+}
	Brown (infusible) when hot, brown when cold.	May be Cd^{2+}
	Red to black when hot, brown when cold.	May be Fe^{3+}, Hg^{2+}
	Green (original colour) to black on heating.	May be Cu^{2+}, Ni^{2+}
	Blue (original colour) to white on heating.	May be Cu^{2+}
	Bright green (original colour) to pale yellow on heating.	May be Ni^{2+}
	Dark pink i.e., crimson (original colour) to blue on heating.	May be Co^{2+}
	Immediate charring i.e., quick blackening from the separation of carbon, smell of burnt sugar.	May be $CH(OH)COO^-$ | $CH(OH)COO^-$

(*Contd...*)

Experiment	Observation	Inference
	Gas evolved	
	Characteristic smell of ammonia, gives dense white fumes with conc. HCl.	May be NH_4^+
	Steam condensed at upper part of the test tube	May be compounds with water of crystallisation.
	Colourless and odourless gas evolves which turns lime water milky.	May be carbonate or bicarbonate of metal.
	Odour of burning sulphur, turns acidified potassium dichromate solution or paper green.	May be sulphite of metal.
	Odour of rotten eggs, turns lead acetate paper or solution black.	May be sulphide of metal.
	Brown, nitrous smell, turns ferrous sulphate solution brownish black.	May be nitrate of heavy metal.
	Reddish brown gas, smell of bromine, turns starch solution orange yellow.	May be bromide in presence of oxidising agent.
	Greenish yellow gas, smell of chlorine, bleaches moist litmus paper.	May be chloride in presence of oxidising agent.
	Violet vapour condensing to black crystals, smell of iodine, turns starch paper blue.	May be iodide in presence of oxidising agent.
	Cracking	
	Cracking Sound	May be oxidising agent like nitrate.
	Sublimate	
	Sublimate formed	May be NH_4^+, Hg_2^{2+}, Sb^{3+}, As^{3+}.

(*Contd...*)

Experiment	Observation		Inference
	(i) A grey sublimate easily convertible to globules when rubbed with a glass rod.		May be Hg_2^{2+}
	(ii) Greyish black sublimate accompanied by a garlic odour.		May be As^{3+}
	(iii) A white sublimate.		May be NH_4^+, Hg_2^{2+}, As^{3+}, Sb^{3+}.
If a sublimate is formed, only then the mixture is mixed with fusion mixture in the ratio of 1:3 and kept in an other test tube and strongly heated.	No sublimate is formed. A gas having characteristic smell of ammonia evolves which turns moist red litmus paper blue, gives dense white fumes when a glass rod moistened with conc. HCl is brought near the mouth of the test tube, gives brown precipitate, or brown or yellow colouration with Nessler's reagent.		May be NH_4^+
	Grey mirror, globules of mercury seen inside the test tube; globules of Hg also obtained when the sublimate is rubbed by a glass rod.		May be Hg_2^{2+}
	Grey sublimate, garlic smell. Does not form globules.		May be As^{3+}
	Black mirror, difficult to obtain.		May be Sb^{3+}
2. Charcoal reducing test:	In the cavity	Outside the cavity	
Mixture of the given salt and fusion mixture in the ratio of 1:3 is kept in the cavity of charcoal block and heated by reducing flame with the help of blow-pipe.	Metallic bead, dull white, brittle, does not mark paper.	Yellow incrustation.	May be Bi^{3+}
	Metallic bead, small in size but large in number, brittle.	White incrustation.	May be Sb^{3+}
	Metallic bead, malleable and marks paper.	Yellow incrustation.	May be Pb^{++}

(*Contd...*)

Experiment	Observation		Inference
	In the cavity	Outside the cavity	
	Red metallic scales.	No incrustation.	May be Cu^{2+}
	Black mass attracted by magnet.	No incrustation.	May be Fe^{3+}
	Black mass, feebly attracted by magnet.	No incrustation.	May be Ni^{2+}, Co^{2+}
	Infusible mass, yellow when hot, white when cold, incandescent when heated.	Yellow when hot, white when cold.	May be Zn^{2+}
	White infusible mass, incandescent when heated.	No incrustation.	May be Ca^{2+}, Sr^{2+}, Ba^{2+}, Al^{3+}, Mg^{2+}

Experiment	Observation	Inference
3. Charcoal oxidising test:		
A pinch of the salt mixture is kept in the cavity of charcoal block and heated strongly by oxidising flame with the help of blow-pipe.	White infusible mass, incandescent when heated.	May be Ca^{2+}, Sr^{2+}, Ba^{2+}, Al^{3+}, Mg^{2+}, BO_3^{3-}, ASO_4^{3-}, PO_4^{3-}, SiO_3^{2-}
	Infusible mass, yellow when hot, white when cold, incandescent when heated.	May be Zn^{2+}
If incandescent mass is obtained, it is moistened with one drop of cobalt nitrate or cobalt chloride solution and again heated strongly for at least 2-3	Blue infusible mass.	May be Al^{3+}
	Blue fusible mass.	May be SiO_3^{2-}, PO_4^{3-}, BO_3^{3-}, AsO_4^{3-}
	Green mass.	May be Zn^{2+}
	Pink mass.	May be Mg^{2+}
	Grey or white mass.	May be Ca^{2+}, Sr^{2+}, Ba^{2+}

(*Contd...*)

Experiment	Observation				Inference
minutes by oxidising flame with the help of blow-pipe.					
4. Flame test: Platinum wire is cleaned. Tip of the wire is moistened with conc. HCl and touched to a particle of the mixture which adheres to it and then heated in a non-luminous Bunsen flame.	Flame colour through naked eye		Flame colour through two blue glasses		
	Voluminous, persistent golden yellow.		No flame colour		May be Na^+
	Violet.		Violet-red		May be K^+
	Brick red (dull red) transient i.e., flickering.		Greenish yellow		May be Ca^{2+}
	Crimson (bright red), permanent.		Purple		May be Sr^{2+}
	Grass green, persistent, seen after strong heating.		Bluish green		May be Ba^{2+}
	Green with blue centre.		No flame colour		May be Cu^{2+}
	Bright green.		—		May be BO_3^{3-}
	Lambent blue.		No flame colour		May be Bi^{3+}, Pb^{2+}, Cd^{2+}, Zn^{2+}
5. Borax bead test: A loop of Pt-wire is made and heated. It is touched to borax and heated to make a bead on the loop. A few particles of mixture is taken on the loop and heated.	Colour in oxidising flame		Colour in reducing flame		
	Hot	Cold	Hot	Cold	
	Green	Blue	Colourless	Opaque red or brown	May be Cu^{2+}
	Yellowish brown	Yellow	Green	Green	May be Fe^{3+}
	Yellow	Green	Green	Green	May be Cr^{3+}
	Blue	Blue	Blue	Blue	May be Co^{2+}

(*Contd...*)

Experiment	Observation				Inference
	Colour in oxidising flame		Colour in reducing flame		
	Hot	Cold	Hot	Cold	
	Violet	Reddish brown	Grey or black	Grey or black and opaque	May be Ni^{2+}
	Reddish violet	Reddish violet	Colourless	Colourless	May be Mn^{2+}

6.1. Explanation of Action of Heat

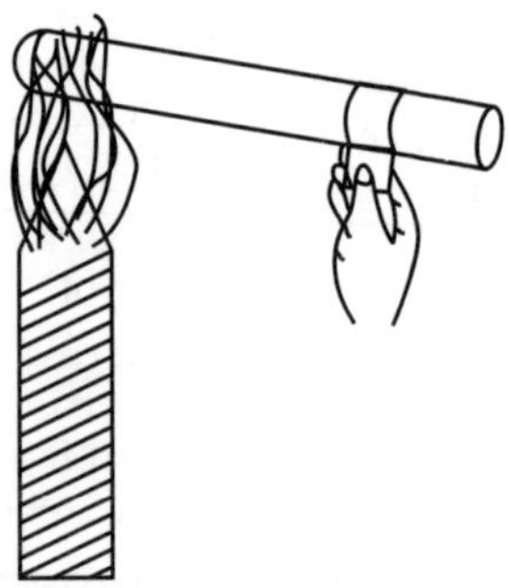

Firstly, a test tube is dried by heating it over Bunsen flame. The mouth of the test tube is kept downwards as shown in the figure so that the condensed vapour may not come back to the hot part. The test tube is kept in rotatory motion while it is being heated. First closed end is heated and then heating is slowly proceeded to the open end.

When the test tube is dried, a pinch of the well powdered mixture is taken in it so that none of it remains adhered to the sides, and heated continuously, at first gently and then strongly. Now the following changes are noted:

(i) Colour change, if any

(ii) Gas evolved, if any

(iii) Cracking, if any

(iv) Sublimate, if any

Colour change and sublimate give indication of basic radicals of the mixture whereas gas evolved and cracking give indication of acid radicals of the mixture.

Change of colour: Some inorganic compounds decompose to their oxides on heating accompanied by changes in colour. The colour of the oxide gives useful information regarding the compound. For example ZnO is yellow while hot and white when cold; PbO is yellow while both hot and cold.

Gas evolved: Carbonates and nitrates of many metals decompose to liberate colourless CO_2 and brown fumes of NO_2, respectively.

Sublimate formed: Ammonium salts and certain salts of Hg as well as As_2O_3 volatilise on being heated and form a white sublimate on the cooler part of the test tube. However, certain salts of Hg and As often give rise to grey and black sublimates, respectively.

Compounds forming white sublimates are mixed with an excess of fusion mixture (Na_2CO_3 + K_2CO_3) and heated resulting in the following changes:

(i) Ammonium salts are converted into ammonium carbonate which decomposes to liberate ammonia gas.

$$2NH_4Cl + Na_2CO_3 \rightarrow (NH_4)_2CO_3 + 2NaCl$$

$$(NH_4)_2CO_3 \rightarrow H_2O + CO_2 + 2NH_3$$

(ii) Mercury salts are likewise transformed into the unstable carbonates which decompose to form the oxides. The oxides then break down to metallic mercury, which forms a grey mirror convertible into globules on rubbing with a glass rod.

$$Hg_2Cl_2 + Na_2CO_3 \rightarrow Hg_2CO_3 + 2NaCl$$

$$Hg_2CO_3 \rightarrow Hg_2O + CO_2$$

$$2Hg_2O \rightarrow 4Hg + O_2$$

(iii) As_2O_3 forms sodium arsenate and metallic arsenic which volatilises with garlic odour and collects as a black mirror on the cooler part of the test tube.

$$5As_2O_3 + 9Na_2CO_3 \rightarrow 6Na_3AsO_4 + 4As + 9CO_2$$

Things to Remember

(i) In case the mixture contains water of crystallisation, steam is evolved during heating and the tube becomes too hot to be held in the hand. In such a case, test tube holder or paper slip should be used to hold the tube. Further, the tube must be held in an almost horizontal position with its mouth slightly downwards. This will prevent the moisture from falling back on the heated bottom and thus cracking it.

(ii) In case of the mixture of several compounds of different colours, change of colour sometimes yields no characteristic information about the constituent of the mixtures.

(iii) Except in certain cases, definite conclusions should not be drawn from negative results, owing to the variety of factors which may prevent the normal results from being obtained. If the test gives a characteristic positive result it is of value, but if the result is negative or indecisive little can be gained from it.

(iv) Equimolecular mixture of sodium carbonate and potassium carbonate is known as fusion mixture.

6.2. Explanation of Charcoal Reducing Test

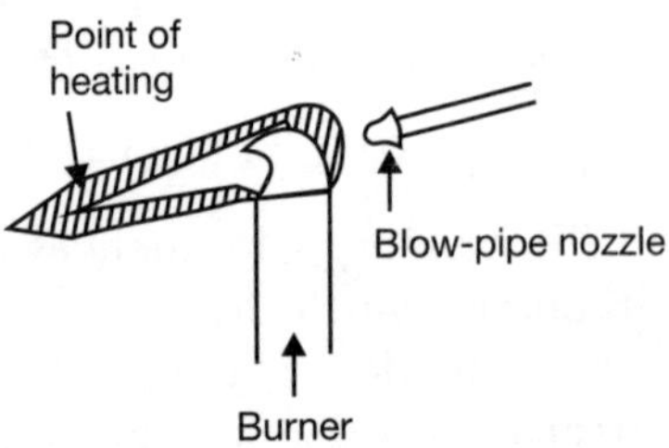

First of all we should know the heating method by reducing flame. The mouth blow-pipe is a simple means by which small and very hot flame can be obtained from a Bunsen flame. The proper use of the blow-pipe can be acquired only by practice, since a steady and continuous blast of air must be maintained for fairly long periods. The mouth blow-pipe is used here to produce a reducing flame.

Air holes of the Bunsen burner are completely closed and the gas supply adjusted to give a luminous flame 4-5 cm in height. Nozzle of the blow-pipe is kept just outside the flame. Air is blown gently and silently from the cheeks. Luminous flame is simply turned. It should graze over the surface of the charcoal block cavity. The material to be heated being placed at the tip of the inner light blue cone.

Now a little of the salt mixture is mixed with fusion mixture in the ratio of 1:3. The mixture is kept in the cavity of charcoal block and heated by reducing flame with the help of blow-pipe. Here the first reaction is the formation of metal carbonate. It decomposes to metal oxide. The oxide may be reduced to metal or remain unchanged.

The oxides of calcium, strontium, barium, aluminium, magnesium are white, infusible and are not reduced by charcoal. They glow very brightly when strongly heated.

The oxides of lead, bismuth, copper, iron, nickel, cobalt are reduced either to a metallic bead (Pb, Bi, Sb) or a sintered mass (Cu) or to a dark metallic powder (Fe, Ni, Co). The oxides of cadmium, arsenic and zinc are readily reduced but as these metals are very volatile, they are easily carried out to the oxidising zone to be converted into difficultly volatile oxides. The oxides thus formed are deposited as incrustation round the cavity.

The fused metal forms the bead on cooling. Any unreduced oxide that volatilises forms incrustation round the cavity. Some oxides which are difficult to be reduced, are left as residue in the cavity.

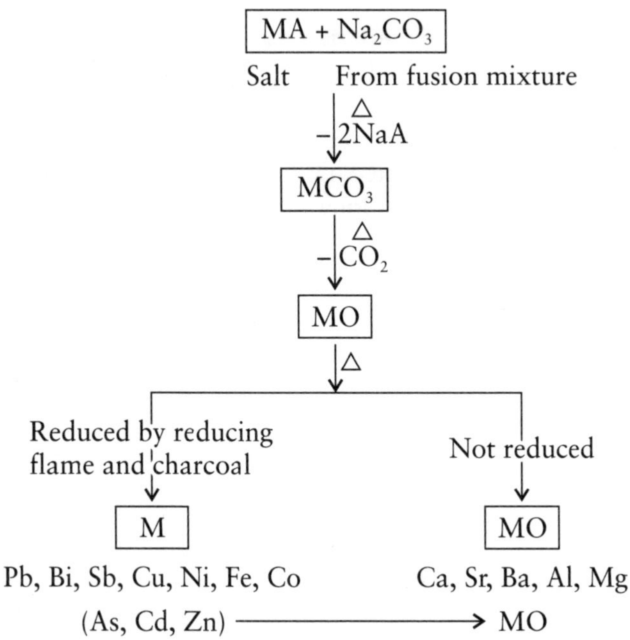

The reactions taking place in this test can be illustrated as under:

$$CuCl_2 + Na_2CO_3 \rightarrow CuCO_3 + 2NaCl$$

$$CuCO_3 \rightarrow CuO$$

$$CuO + C \rightarrow Cu$$

Things to Remember

(i) An incrustation is formed on the cooler portion of the charcoal, surrounding the cavity.

(ii) With ordinary charcoal block it is difficult to get incrustation. Analytically prepared blocks are preferable to natural wood charcoal.

(iii) A substance is said to be incandescent when it emits light like a glowing mantle.

6.3. Explanation of Charcoal Oxidising Test

For heating by oxidising flame air holes of the Bunsen burner are closed and nozzle of the blow-pipe is placed just inside the flame. Air is blown in rather hard from the cheeks. The material to be heated is placed at the tip of the long pointed flame.

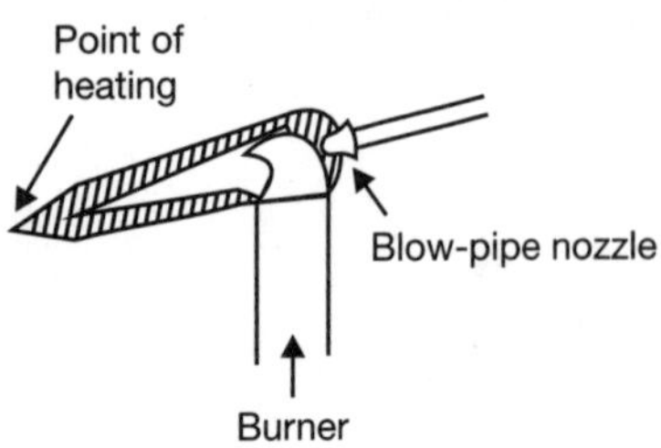

Now a cavity is scooped out in a charcoal block. A pinch of salt mixture is kept in the cavity and heated strongly by oxidising flame with the help of blow-pipe. Salts when heated strongly in air are converted into oxides. The oxide when moistened with cobalt nitrate or cobalt chloride solution and heated strongly by oxidising flame, aluminium, zinc, magnesium, phosphate, borate, etc. yield coloured mass. Cobalt nitrate or cobalt chloride is itself reduced to cobalt oxide. This oxide combines with metallic oxide (under investigation) to form a "double oxide" which has a characteristic colour.

Thus the reactions taking place in the cavity of the charcoal block can be depicted as under:

$$2ZnS + \underset{\text{(From air)}}{3O_2} \rightarrow 2ZnO + 2SO_2 \uparrow$$

$$2Co(NO_3)_2 \rightarrow 2CoO + 4NO_2 \uparrow + O_2 \uparrow$$

$$ZnO + CoO \rightarrow \underset{\substack{\text{(Cobalt zincate;} \\ \text{green mass} \\ \text{i.e. Rinmann's green)}}}{ZnO \cdot CoO}$$

Similarly,

$$CoO + Al_2O_3 \rightarrow \underset{\substack{\text{(Cobalt meta aluminate;} \\ \text{blue mass, i.e., Thenard's blue)}}}{Co(AlO_2)_2}$$

$$CoO + MgO \rightarrow \underset{\text{(Pink mass)}}{CoO \cdot MgO}$$

In case of phosphate, borate, arsenate and silicate the corresponding cobalt salt is produced which has a blue colour.

Things to Remember

(i) Only a drop of cobalt nitrate or cobalt chloride solution is to be used for performing the test otherwise CoO obtained from the decomposition of cobalt salt turns the residue black instead of giving the characteristic colour.

(ii) Sometimes complementary colours are produced due to the combination of two colours formed during the heating on charcoal.

(iii) Indication for aluminium cannot be conclusive unless the borate, silicate, phosphate and arsenate radicals have been shown to be absent.

6.4. Explanation of Flame Test

Flame test is based upon the fact that certain volatile salts impart characteristic colours to the non-luminous flame. The metallic chlorides are amongst the most volatile compounds and when taken to the non-luminous flame of Bunsen burner, they impart characteristic flame colours.

Metallic chlorides are prepared by moistening the salt to be tested with little concentrated hydrochloric acid before carrying out the test.

On observing the flame through blue glass, a part of the coloured light is absorbed by the glass and so flame colouration undergoes a characteristic change.

Platinum wire is used as the medium for carrying the chloride paste to the flame because platinum, one of the most "noble" metals, does not undergo chemical change.

To perform the test, firstly the tip of the wire is heated in the non-luminous Bunsen flame using the hottest part of the flame. If any colour is imparted to the flame on heating the wire, owing to the presence of traces of impurity on the surface, the tip of the wire is dipped into a little concentrated hydrochloric acid kept in a watch glass and re-heated, sufficient repetition of this procedure should give a clean wire which does not colour the Bunsen flame. Now the wire is moistened with concentrated hydrochloric acid and brought into contact with a small particle of the mixture which adheres to the tip of the wire. Now the wire is carefully introduced to the lower oxidising flame and colour of the flame is noted through the naked eye and also through the blue glass. If no coloured flame is observed then the wire is dipped in conc. HCl and reheated strongly. The test should be repeated 2-3 times and only then the result should be declared "No characteristic flame colour".

Things to Remember

(i) Concentrated hydrochloric acid for flame test should be kept in watch glass. Contaminated acid should not be used. Platinum wire should never be dipped in the bottle of the acid.

(ii) In case of easily reducible metals (for example Cu, Pb, Sb, Bi, As, Sn, etc.) where we get metallic bead on charcoal, the platinum wire should not be used as the metal forms alloy with it and makes it brittle. In such cases, tip of black lead-pencil may be used.

(iii) It is necessary to heat the wire alone and then it should be moistened with conc. HCl. The salt is taken only once. Then it is heated for a couple of minutes before any desired result is expected. By alternate moistening with conc.

HCl and heating, the substance gradually changes over to the volatile chloride and then gives the characteristic flame colour.

(iv) Many salts which contain mere traces of sodium as impurity give "sodium flame". Thus students experience difficulty to decide. Experience shows that small amount of sodium gives only a transient flame colouration, while larger quantity of sodium gives a strong persistent golden yellow flame. It is observed that the yellow flame of sodium masks that of other elements, e.g., of potassium. In this case two blue glasses (cobalt) are jointly used when the flame is observed through blue glasses, the yellow colour of the sodium flame is absorbed and other colours get modified, e.g., that of potassium looks violet-red instead of violet.

(v) Salts of barium are rather reluctant to give the flame test. Therefore, the platinum wire with salt should be kept in the flame for about 4 minutes and only then declared "No indication". With salts like barium sulphate or phosphate which are not converted into chloride easily the green flame of barium is either indistinct or not visible.

(vi) Potassium salts volatilise at low temperature. Hence presence of potassium salts may be shown by carefully introducing the salt into the cooler portion of the flame at the base.

(vii) Flame of strontium is rich in volume and lasts long and as such there is no difficulty in detecting strontium but in case of calcium, when strontium is also present, it is difficult to detect calcium. Calcium chloride is much more volatile than strontium chloride and this is utilised in detecting calcium in presence of strontium. Platinum wire is very carefully introduced in the flame and colour is observed just at the time of introducing the wire.

(viii) Sometimes on the platinum wire there is a deposit of salts which is not removed by dipping in conc. hydrochloric acid and heating. In this situation, the hot wire is touched with solid potassium hydrogen sulphate so as to produce a coating of the salt on this wire. On heating the wire gently a bead of potassium pyrosulphate is formed which travels along the wire dissolving all deposited salt. On cooling, the bead is easily removed. Any pyrosulphate adhering dissolves in water and finally the wire is cleared by usual dipping in conc. HCl and heating.

(ix) The student should carefully distinguish between the flames of barium, borate and copper. The flame of barium is grass green and persistent, the flame of borate is bright green but has no blue centre whereas the flame of copper is bright green with blue centre.

6.5. Explanation of Borax Bead Test

This test is based on the fact that fused borax reacts with compounds of certain metals to form borates which have characteristic colours of the metallic ions. Some metals form two borates; one derived from the higher oxide which is formed in the oxidising flame and the other from the lower oxide which is formed in the reducing flame. This is why both oxidising and reducing flames are used in the borax bead test.

Ordinary borax is sodium tetraborate decahydrate ($Na_2B_4O_7 \cdot 10H_2O$). On heating, borax swells owing to the loss of water of crystallisation. The anhydrous borax so produced may be looked upon as built up of sodium metaborate and boric anhydrides, both of which are fusible and transparent.

$$Na_2B_4O_7 \cdot 10H_2O \rightarrow Na_2B_4O_7 + \underset{\text{(Vap.)}}{10H_2O} \uparrow$$

$$Na_2B_4O_7 \rightarrow \underset{\text{(Sod. metaborate)}}{2NaBO_2} + \underset{\text{(Boric anhydride)}}{B_2O_3}$$

When the borax bead is brought into contact with a coloured salt and heated in oxidising flame, the latter dissolves in boric anhydride to form a coloured metaborate.

$$B_2O_3 + CuSO_4 \rightarrow \underset{\text{(Cupric metaborate; blue)}}{Cu(BO_2)_2} + SO_3 \uparrow$$

When the coloured bead is heated in the reducing flame, the carbon of the flame reduces the higher metaborate to a lower one, thus bringing about a profound change in the colour. Sometimes, the reduction results in the formation of free metals (Ni and Cu salts).

$$\underset{\text{(Cupric metaborate; blue)}}{2\,Cu(BO_2)_2} + C \rightarrow \underset{\text{(Cuprous metaborate; colourless)}}{2\,CuBO_2} + B_2O_3 + CO$$

$$2\,CuBO_2 + C \rightarrow \underset{\text{(Metal; opaque red)}}{2Cu} + B_2O_3 + CO$$

To perform the borax bead test, a small loop is made at the end of the platinum wire as depicted below:

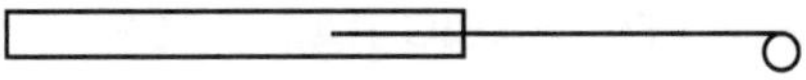

The platinum loop is heated to bright red in the non-luminous flame of Bunsen burner and dipped quickly into powdered borax, some of the borax should adhere to the wire. On heating the loop in hottest part of the flame, borax swells and then fuses to a clear and colourless bead, which should just fill the loop, if it does not do so, the process should be repeated. Now only a few particles of the salt mixture is taken on the loop and heated strongly in the oxidising zone(C) of non-luminous flame. The loop is now taken out and the colour of the bead noted both when it is hot and cold. Now the same bead is heated in the lower reducing zone(B) of the non-luminous flame. The bead is then cooled in the zone of unburnt gas(A) in order to prevent oxidation, which might take place if the bead is cooled in air. Now the bead is taken out and the colour noted both when it is hot and cold.

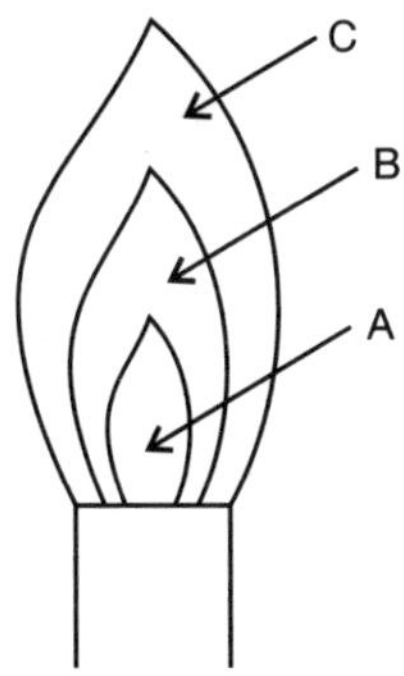

Non-luminous flame

(C) Oxidising zone
(B) Reducing zone
(A) Zone of unburnt gas

In order to clean the platinum wire, it is heated strongly in oxidising zone of non-luminous flame. When the loop and the bead are red hot, a jerk is given to the glass tube with the left hand. The bead falls down and the wire is clean for next operation.

Things to Remember

(i) If the bead is black or very dark in colour, it indicates that too much salt has been used. In this situation, fresh bead should be prepared.

(ii) If the salt mixture is heterogeneous in character, different particles of different colours should be separated mechanically and the borax bead tests should be performed separately for different particles of different colours.

7. WET TESTS FOR BASIC RADICALS

Wet tests are those tests which are applied to the solution of the salt mixture. Dry tests are only preliminary tests. On the other hand, wet tests are comprehensive and conclusive. Sometimes, the student may be tempted to proceed to the wet tests directly. He is advised not to fall prey to such temptation.

7.1. Preparation of Solution

First requirement of wet tests is the preparation of solution of salt mixture in suitable solvent. Now the first problem arises about choice of solvent. What should the properties of ideal solvent be? Which solvent should be selected to prepare the solution? The ideal solvents may have following properties:

(i) Ideal solvent should have capability of dissolving a wide range of substances, so that only a few will require special treatment.

(ii) They should react with only a few substances, so that there be a minimum disturbance of the substance to be analysed.

(iii) They must be plentiful and cheap.

Fortunately, water comes near to fulfilling all these requirements. Tap water is unfit for wet tests. It always contains bicarbonates, chlorides and sulphates of calcium and magnesium. Distilled water is essential for accurate work and its use is intended whenever water is mentioned in the following pages.

Next to water, three mineral acids viz. hydrochloric acid, nitric acid and sulphuric acid come for selection. Sulphuric acid may immediately be discarded because it forms insoluble sulphates (Ca, Sr, Ba, etc.). It is an oxidising agent, hence in Group II, it will produce enormous amount of sulphur.

All metallic nitrates are soluble in water, hence nitric acid may be used as a solvent. But this too should be avoided as it is an oxidising agent. It will produce enormous amount of sulphur in Group II.

Hydrochloric acid is free from these difficulties. Salts which are not soluble in water are usually soluble in hydrochloric acid except Pb^{2+}, Hg^{+} and Ag^{+} which are soluble in dil. HNO_3. Thus,

(i) First solvent is distilled water

(ii) Second solvent is hydrochloric acid

(iii) Third solvent is nitric acid

To prepare the solution of the given salt mixture, following list is being given in order of preference:

(i) Distilled water
- → Shake at room temperature.
- → Heat to boiling and then allow to cool to room temperature.

(ii) Dilute HCl
- → Shake at room temperature.
- → Heat to boiling and then allow to cool to room temperature.

(iii) Mod. conc. HCl
- → Shake at room temperature.
- → Heat to boiling and then allow to cool to room temperature.

(iv) Dil. HNO_3
- → Shake at room temperature.
- → Heat to boiling and then allow to cool to room temperature.

To prepare the solution of the mixture, a pinch of it is taken in a test tube and 10 c.c. solvent is added to it. The test tube is shaken for dissolution. If it does not dissolve, the test tube is heated to boiling with shaking. This procedure is adopted by taking solvent one by one in the order mentioned above until the mixture dissolves.

After knowing the solubility of the salt mixture, 1-2 g of it is taken in a beaker. About 100 c.c. appropriate solvent is added and the salt mixture dissolved. This will be the stock solution which will be used wherever it will be necessary.

Things to Remember

(i) If dilute nitric acid is used as solvent to prepare the solution, it should be kept in mind that metals of the salts have been converted into their nitrates.

Salt + HNO_3 $\rightarrow$ Metal nitrate

The excess nitric acid present in the solution must be removed, otherwise it will produce enormous amount of sulphur in Group II due to its oxidising nature. Excess of nitric acid is removed by evaporating the solution in a basin nearly to dryness. It is diluted with water. Now the solution is ready.

(ii) Before proceeding to 'group separation', one should have idea about the presence of interfering radicals. This is why tests for acid radicals are done before those for basic radicals. If interfering radical is present, it must be removed either before the preparation of solution or at proper place in 'group separation'.

7.2. Knowledge of Groups and Group Reagents

Wet tests for basic radicals are very systematic. Basic radicals have been divided into several groups on the basis of the reagents, called group reagents, employed for the precipitation of metals as their insoluble salts.

Groups	Metal ions	Group reagents
I	Hg_2^{2+}, Ag^+, Pb^{2+}	Dil. HCl
II	Pb^{2+}, Hg^{2+}, Bi^{3+}, Cu^{2+} Cd^{2+}, As^{3+}, Sn^{4+}, Sb^{3+}	Dil. HCl + H_2S
III	Fe^{3+}, Al^{3+}, Cr^{3+}	NH_4Cl + NH_4OH

(Contd...)

Groups	Metal ions	Group reagents
IV	Ni^{2+}, Co^{2+}, Mn^{2+}, Zn^{2+}	$NH_4Cl + NH_4OH + H_2S$
V	Ca^{2+}, Sr^{2+}, Ba^{2+}	$NH_4Cl + NH_4OH + (NH_4)_2CO_3$
VI	Mg^{2+}, Na^+, K^+, NH_4^+	No particular group reagent

7.3. Removal of Interfering Radicals

Interfering radicals are those acid radicals which interfere in general course of group separation, i.e., separation of metal ions into groups.

Oxidising ions like permanganate, dichromate, chromate, etc. interfere in separation of Group II. As hydrogen sulphide gas is passed in group II, enormous amount of sulphur is formed.

Organic acids like oxalate, tartarate, etc. and phosphate, borate, silicate, fluoride, etc. interfere in separation of Group III.

7.3.1. Removal of oxidising ions

About 2 g of the original salt mixture is taken in a basin and about 10 c.c. conc. HCl is added to make a thin paste. It is heated strongly till the reduction is complete (indicated by the change of colour). It is cooled and diluted with water. It is filtered if necessary.

$$\left[MnO_4^- + 8H^+ + 5e \rightarrow Mn^{2+} + 4H_2O\right] \times 2$$

$$\left[2Cl^- \rightarrow Cl_2 + 2e\right] \times 5$$

$$2MnO_4^- + 10Cl^- + 16H^+ \rightarrow 2Mn^{2+} + 8H_2O + 5Cl_2 \uparrow$$

Things to Remember

(i) Addition of either 5 c.c. rectified spirit or 2-3 c.c. of 10% formaldehyde solution to the HCl paste helps the reduction.

(ii) If Group I is absent, removal of oxidising ion is done from the original salt mixture.

(iii) If Group I is present, removal of oxidising ion is done from the filtrate of Group I. In this case, filtrate of Group I is heated to dryness and then removal of oxidising ion is done from this dry mass.

(iv) If arsenate is suspected then the hydrochloric acid paste after strong heating should not be diluted much. In this stage, arsenate is reduced into arsenite. Further, for precipitation of As^{3+} as As_2S_3 in Group II, H_2S gas should be passed into hot and concentrated solution for atleast 5-7 minutes.

7.3.2. Removal of organic acids

The filtrate from Group II is evaporated to dryness, when some carbon may be liberated by the decomposition of organic acids. By repeated evaporation with conc. HNO_3, the black residue is completely oxidised and no free nitric acid remains. The residue must not be heated too strongly as Fe_2O_3, Cr_2O_3 and Al_2O_3 so formed may be rendered very sparingly soluble in HCl. Lastly, the mass left is then treated with dil. HCl and proceeded for separation of Group III.

Things to Remember

(i) If cations of Groups I and II are absent, removal of organic acid is done from original salt mixture.

(ii) If cations of Groups I and II are present, removal is done from the filtrate of Group II.

7.3.3. Removal of borate, silicate and fluoride

Group II filtrate is taken in a basin and evaporated to dryness. About 2-3 c.c. conc. HCl is added and evaporated to dryness. This process is repeated 2-3 times. The HF volatilises with the HCl and the H_3BO_3 volatilises in the steam.

$$F^- + H^+ \rightarrow HF$$

$$BO_3^{3-} + 3H^+ \rightarrow H_3BO_3$$

Silicate is converted into silica.

$$SiO_3^{2-} + 2H^+ \rightarrow H_2SiO_3$$

$$H_2SiO_3 \rightarrow SiO_2 \downarrow + H_2O$$

The residual mass is extracted with dilute HCl and the SiO_2 is filtered off. Then we can proceed for separation of Group III from the filtrate.

Things to Remember

(i) If cations of Groups I and II are absent, removal can be done from original salt mixture.

(ii) If cations of Groups I and II are present, removal is done from the filtrate of Group II.

7.3.4. Removal of phosphate

There are several methods for the removal of phosphate ion. Of these, three methods will be outlined here. The first one, the zirconium nitrate method is the most convenient; if correctly performed. All phosphate ions are quantitatively and rapidly removed as the zirconium phosphate is highly insoluble. The second is the iron(III)-acetate method. This method has been extensively used in the past. It can be adapted if zirconium nitrate is not available in the laboratory. Finally, the tin(IV) chloride method is also described because this method can be applied in presence of the 'rarer' elements (Ce, V, etc.).

Students are advised to follow:

Table 4 for zirconium nitrate method,

Table 5 for Iron(III)-acetate method and

Table 6 for tin(IV) chloride method.

Table 4: Zirconium Nitrate Method

Volume of filtrate from Group II is reduced to about 10 c.c. by evaporation, if necessary. The HCl concentration should not exceed M. About 0.5 - 1 g solid NH_4Cl is added and stirred until dissolved. Then the reagent, zirconium (IV) nitrate is slowly added with stirring until precipitation is complete. A large excess of the reagent must be avoided. Now the content of the test tube or small conical flask is heated just to boiling and stirred with a glass rod to prevent bumping. It is filtered through Whatman No. 32 filter paper. The precipitate is washed with a little hot water and the washings combined with the filtrate.	
Residue:	Filtrate:
Zirconium Phosphate Rejected	Presence of phosphate is tested by addition of a drop of zirconium(IV) nitrate. About 0.5 g solid NH_4Cl is added and heated to boiling. Then slight excess of dil. NH_4OH is added (confirm by smell of NH_3). It is boiled for 2-3 minutes and filtered.

(Contd...)

	Residue:	Filtrate:
	The residue is subjected for analysis of Group III. The excess of zirconium will be found in the residue after treatment with H_2O_2 and NaOH solution (or with sodium peroxoborate, $NaBO_3 \cdot 4H_2O$) and boiling. It will accompany $Fe(OH)_3$ if iron is present.	The filtrate is subjected for separation of Group IV onwards.

Things to Remember

(i) The HCl concentration should not exceed M, otherwise a turbid supernatant liquid is obtained and the precipitation of phosphate is not quantitative.

(ii) It is important that the excess of the zirconium nitrate reagent should not exceed 25 per cent, otherwise a turbid supernatant liquid will be obtained. This turbidity cannot be removed by filtration or centrifugation.

Table 5: Iron(III)-acetate Method

The filtrate from Group II is heated to expel H_2S. About 1 c.c. conc. HNO_3 is added and heated to convert Fe^{2+} ion into Fe^{3+} ion, if Fe^{2+} ion is present. Now small portion of the solution is taken and potassium ferrocyanide or ammonium thiocynate solution is added to it. If iron is present, potassium ferrocyanide will produce blue precipitate and ammonium thiocynate will produce red colouration. Now to the main volume of the cold solution, dil. NH_4OH is added dropwise with stirring until either a faint permanent precipitate is just obtained or until the solution is just alkaline (test with litmus paper). Then 2-3 c.c. dil. CH_3COOH (1:1) and 5 c.c. 10M ammonium acetate are added. If the solution is red or brownish red, there is no need of addition of neutral $FeCl_3$ solution. If the solution is not brownish red, neutral $FeCl_3$ solution is added dropwise with shaking until the solution acquires deep brownish red colouration. It is diluted and boiled for 1-2 minutes. The solution is now filtered hot. Residue is washed with little boiling water.	
Residue:	Filtrate:
May contain phosphate and basic acetate of Fe, Al, Cr and also $Fe(OH)_3$.	The filtrate is boiled to evaporate to 20-25 ml. About 0.5 g NH_4Cl and dil. NH_4OH (in slight excess)

(*Contd...*)

<table>
<tr><td colspan="2" rowspan="3">Group III Present:
The precipitate is transferred to porcelain dish by means of 10 c.c. cold water. About 1-1.5 g sodium peroxoborate, $NaBO_3 \cdot 4H_2O$ (or 5 c.c. NaOH and 5 c.c. 3% H_2O_2 or 5-10 ml dil. HCl and then Na_2O_2 in small quantities at a time, until the solution becomes strongly alkaline) is added and boiled gently until the evolution of O_2 ceases (2-3 minutes). Now it is filtered and washed with a little hot water.</td><td colspan="2">are added to it. It is filtered, if necessary.</td></tr>
<tr><td>Residue:</td><td>Filtrate:</td></tr>
<tr><td rowspan="3">The residue is examined for Al^{3+} and Cr^{3+}, if not previously tested for.
In general, no precipitate will be obtained here.</td><td rowspan="3">The filtrate is subjected for separation of Group IV onwards.</td></tr>
<tr><td>Residue:</td><td>Filtrate:</td></tr>
<tr><td>$FePO_4$,
$Fe(OH)_3$
Rejected</td><td>May contain $[Al(OH)_4]^-$ and CrO_4^{2-}.
The filtrate is subjected to test for Al^{3+} and Cr^{3+}.</td></tr>
</table>

Things to Remember

(i) $FePO_4$, $AlPO_4$ and, to a lesser extent, $CrPO_4$ are insoluble in warm dilute acetic acid buffered by a solution of ammonium acetate; the phosphates of the metals of Groups IV and V and of Mg are soluble.

(ii) The precipitation of phosphate ions will be complete only if the trivalent ions are present in excess. The procedure depends upon having excess iron(III) ions present. This ion is chosen because it is easy to test for Fe^{3+} in the filtrate from Group II, and iron(III) phosphate is the least soluble in the acetic acid medium. Furthermore, the excess of iron(III) ions can be readily removed since iron(III) acetate is considerably hydrolysed upon dilution and boiling, and the excess iron is precipitated as basic iron(III) acetate or as iron(III) hydroxide.

$$Fe^{3+} + 3CH_3COO^- + 2H_2O \rightleftharpoons Fe(OH)_2CH_3COO \downarrow + 2CH_3COOH$$

$$Fe^{3+} + 3CH_3COO^- + 3H_2O \rightleftharpoons Fe(OH)_3 \downarrow + 3CH_3COOH$$

The reaction tends to reverse upon cooling, hence filtration of the hot solution is desirable.

(iii) The addition of ammonium acetate solution to a solution containing iron, aluminium, and chromium ions yields the acetate of these metals. Upon boiling the highly diluted solutions, the basic acetate of iron is almost completely and that of aluminium is largely precipitated. Chromium acetate does not appear to form a basic acetate, but in the presence of considerable amount of iron and aluminium, chromium is coprecipitated as basic acetate.

(iv) Ordinary $FeCl_3$ solution contains HCl in it, which is produced by the hydrolysis of the salt. Hydrochloric acid, in phosphate removal, prevents the precipitation of the basic acetates. Therefore, neutral ferric chloride is emloyed. This is prepared by adding dilute ammonium hydroxide solution dropwise until a slight permanent precipitate is formed. The precipitate is filtered off and the filtrate is now called neutral ferric chloride solution.

(v) Presence of iron must be tested before the removal of phosphate as ferric chloride is added in the process. Potassium ferrocyanide gives blue precipitate and ammonium thiocyanate gives blood red colouration.

$$4Fe^{3+} + 3\left[Fe^{II}(CN)_6\right]^{4-} \rightarrow \underset{\text{(Ferric ferrocyanide; prussian blue)}}{Fe_4^{III}\left[Fe^{II}(CN)_6\right]_3}$$

$$Fe^{3+} + 6SCN^- \rightarrow \underset{\text{(Ferric thiocynate ion; blood red)}}{\left[Fe^{III}(SCN)_6\right]^{3-}}$$

Table 6: Tin(IV) Chloride Method

The filtrate from Group II is boiled to expel H_2S and diluted to 50 c.c. About 3 c.c. 0.5M $(NH_4)_2HPO_4$ solution (if necessary) is added followed by dilute NH_4OH until the precipitate which forms initially does not redissolve on shaking. Now 1 c.c. dil. HCl is added and heated to boiling. Then 2 c.c. of the $SnCl_4$ reagent is introduced. It is filtered by Whatman filter paper. Test of phosphate must be done from filtrate.

(*Contd...*)

<table>
<tr><td>Residue:</td><td colspan="2">Filtrate:</td></tr>
<tr><td rowspan="3">It is examined for Cr^{3+} and rejected.</td><td colspan="2">It is treated with H_2S till complete precipitation of SnS_2. It is filtered.</td></tr>
<tr><td>Residue:</td><td>Filtrate:</td></tr>
<tr><td>Rejected</td><td>It is concentrated by boiling with HCl and subjected for separation of Group III onwards.</td></tr>
</table>

Things to Remember

(i) Phosphate is removed as insoluble tin(IV) phosphate, $Sn_3(PO_4)_4$.

(ii) When small amount of phosphate is present, the precipitate filters with difficulty and the filtrate may be turbid. In such case the addition of more phosphate (diammonium phosphate) to the solution before precipitation with the reagent is recommended.

(iii) If the $Sn_3(PO_4)_4$ precipitate is green in colour, Cr might be present. To test for this small portion of the precipitate is boiled with 2M NaOH solution. It is cooled, filtered. Now 3 per cent H_2O_2 and 1-2 c.c. amyl alcohol is added to the filtrate. A blue colouration of the amyl alcohol layer confirms Cr.

7.4. Group Separation

For separation of metal ions into groups, following Table 7 must be followed.

Table 7

<table>
<tr><td colspan="3">Dil. HCl is added to cold solution of salt mixture till complete precipitation, if precipitation takes place. It is filtered.</td></tr>
<tr><td>Residue:</td><td colspan="2">Filtrate:</td></tr>
<tr><td rowspan="3">White. Group I Present</td><td colspan="2">Oxidising interfering radical is removed (if present) as discussed before. Now H_2S is passed till complete precipitation and filtered.</td></tr>
<tr><td>Residue:</td><td>Filtrate:</td></tr>
<tr><td>Coloured. Group II Present</td><td>The filtrate is boiled to expel H_2S. About 1-2 c.c. conc. HNO_3 is added and boiled. Non-oxidising interfering radical is removed (if</td></tr>
</table>

(Contd...)

<table>
<tr><td rowspan="7"></td><td rowspan="7"></td><td colspan="4">present). Now about 1-2 g solid NH_4Cl is added. It is boiled and then slight excess dil. NH_4OH is added till the odour of ammonia persists. It is filtered immediately.</td></tr>
<tr><td>Residue:</td><td colspan="3">Filtrate:</td></tr>
<tr><td rowspan="5">Group III Present</td><td colspan="3">About 2-3 c.c. dil. NH_4OH is added, boiled and then H_2S is passed for 1 minute under pressure. It is filtered.</td></tr>
<tr><td>Residue:</td><td colspan="2">Filtrate:</td></tr>
<tr><td rowspan="3">Group IV Present</td><td colspan="2">The filtrate is transferred to a porcelain dish and acidified with dil. acetic acid immediately. It is evaporated to a pasty mass and 3-4 c.c. conc. HNO_3 is added so as to wash the solid round the wall to the centre of the dish. It is heated cautiously until the mixture is dry. Again it is heated strongly to volatilise the ammonium salt. After cooling, dil. HCl and 10 c.c. water are added. Now solid NH_4Cl and dil. NH_4OH are added till it becomes alkaline. Then $(NH_4)_2 CO_3$ solution is added in excess. It is warmed for 4-5 minutes and filtered.</td></tr>
<tr><td>Residue:</td><td>Filtrate:</td></tr>
<tr><td>White. Group V Present</td><td>It is evaporated to a pasty mass and then 3 c.c. conc. HNO_3 is added and further heated to dry mass and until white fumes of ammonium salt ceases. The white residue is subjected for analysis of Group VI.</td></tr>
</table>

Things to Remember

(i) Larger quantities of the substances must be avoided to save time in filtering, washing and dissolving.

(ii) If the original mixture is completely soluble in dil. HCl, it is evident that no silver or mercurous salt is present. When lead is present, the solution may be clear while hot but $PbCl_2$ is precipitated as fine colourless needles on cooling. So it is always desirable to bring the solution to ordinary temperature before adding dil. hydrochloric acid for precipitation of Group I metals. If lead is present in very small quantity, it may not be precipitated in Group I but it will be precipitated in Group II as sulphide.

(iii) A precipitate may form on addition of dil. HCl to certain neutral or slight acid solution even when none of Group I metals is present. This may occur under the following conditions:

- Aqueous solution of Sb, Bi and Sn, not containing free HCl, precipitate as the oxychlorides upon the addition of the hydrochloric acid. The precipitate dissolves, however, upon the addition of excess acid:

  $$Sb^{3+} + Cl^- + H_2O \rightarrow SbOCl \downarrow + 2H^+$$

  $$Bi^{3+} + Cl^- + H_2O \rightarrow BiOCl \downarrow + 2H^+$$

  $$Sn^{2+} + Cl^- + H_2O \rightarrow Sn(OH)Cl \downarrow + H^+$$

- Concentrated solutions of certain chlorides, e.g. NaCl and $BaCl_2$, may form precipitates upon the addition of HCl; these dissolve on diluting with water.
- Borates may yield a white crystalline precipitate of boric acid, particularly if the acid is concentrated; only partial precipitation may occur here.

  $$BO_3^{3-} + 3H^+ \rightarrow H_3BO_3 \downarrow$$

- Silicates may yield a gelatinous precipitate of silicic acid; only partial precipitation may occur here.

  $$SiO_3^{2-} + 2H^+ \rightarrow H_2SiO_3 \downarrow$$

- Thiosulphates may give white precipitate of sulphur with HCl.

$$S_2O_3^{2-} + 2H^+ \rightarrow S \downarrow + H_2O + SO_2 \uparrow$$

(iv) Lead is not completely precipitated as chloride in Group I. It can be explained as under:

- The value of solubility product of $PbCl_2$ is comparatively much higher than those of AgCl and Hg_2Cl_2.
- Pb^{2+} ion partly remains as $[Pb(OH)]^+$.

$$Pb^{2+} + H_2O \rightarrow \underset{\text{Hydroxo lead (II) ion}}{[Pb(OH)]^+} + H^+$$

(v) Addition of large excess of dil. HCl for precipitation of metals as chlorides in Group I is prohibited as precipitated AgCl and $PbCl_2$ go in solution forming complexes.

$$AgCl + Cl^- \rightarrow \underset{\text{Dichloro argentate(I) ion}}{[AgCl_2]^-}$$

$$PbCl_2 + 2Cl^- \rightarrow \underset{\text{Tetrachloro plumbate(II) ion}}{[PbCl_4]^{2-}}$$

(vi) Student should have own delivery tube and it should be washed thoroughly each time under running tap water.

(vii) In Group II, H_2S should be passed till complete precipitation. It should be tested by passing H_2S into the filtrate obtained from Group II.

(viii) If oxidising ion is present, it must be removed before passing H_2S in Group II, otherwise enormous amount of sulphur will be precipitated.

(ix) After boiling the filtrate (to expel H_2S) obtained from Group II, it is heated with 2-3 c.c. conc. HNO_3 to convert Fe^{2+} ion to Fe^{3+} ion (if Fe^{2+} is present). This is so because iron is precipitated in Group III as ferric hydroxide. Thereafter non-oxidising interfering radical must be removed (if present) before addition of solid NH_4Cl and NH_4OH as phosphates of metals of Groups III, IV, V and Mg are insoluble in alkaline medium.

(x) Before the removal of phosphate (if present) by iron(III)-acetate method, Fe^{3+} ion must be tested by adding potassium ferrocyanide solution or ammonium thiocyanate solution to small portion of filtrate obtained from Group II; former will give blue precipitate and latter will give blood red colouration. It is so because neutral ferric chloride is added in this method.

(xi) To test the complete precipitation in Group III, a little NH_4OH should be added to a small portion of filtrate from group III.

(xii) To test the presence of metals of Group IV, H_2S should be passed into a small portion of the filtrate from Group III. If metals of IV are absent, there is no need to pass H_2S into whole filtrate obtained from Group III.

(xiii) The filtrate from Group IV must be acidified immediately for following reasons:

- The filtrate contains ammonium sulphide which oxidises to ammonium sulphate on exposure to air (slowly). This causes precipitation of $BaSO_4$ and $CaSO_4$.
- On absorption of CO_2 from air produces carbonate which causes precipitation of carbonates of Group V metals.

(xiv) The filtrate from Group IV will contain a very high concentration of ammonium salts. It may lead to incomplete precipitation of the carbonates of Group V metals. For this reason most of the ammonium salts should be eliminated first. This can either be done by simple volatilization or by adding conc. HNO_3.

$$NH_4^+ + HNO_3 \rightarrow N_2O \uparrow + H^+ + 2H_2O$$

(xv) The ammonium carbonate solution contains much ammonium hydrogen carbonate. It leads to the formation of soluble bicarbonate of Ca, Sr and Ba. Thus excess ammonium carbonate should be added. Moreover, gentle heating (50°C) is recommended after addition of ammonium carbonate for decomposition of bicarbonate into carbonate.

7.5. Group Analysis

Analysis of groups for identification of metal ions have been tabulated below:

Table 8: Analysis of Group I

<table>
<tr><td colspan="3">The ppt. may contain $PbCl_2$, AgCl and Hg_2Cl_2. The ppt. on the filter paper is washed first with dil. HCl and then with small amount of cold water. It is transferred to a hard glass test tube and boiled with 25 c.c. water. Then it is filtered hot.</td></tr>
<tr><td colspan="2">Residue:</td><td>Filtrate:</td></tr>
<tr><td colspan="2">It may contain Hg_2Cl_2 and AgCl. The ppt. is washed with hot water until the washings give no ppt. with potassium chromate solution (test of lead). About 15 c.c. of warm NH_4OH solution (1:1) is added over the precipitate on filter paper and the filtrate collected.</td><td rowspan="3">It may contain $PbCl_2$. The filtrate is divided into two parts.
Part I—On cooling under running tap water with jerks, white crystalline ppt. is obtained.
Pb^{2+} present.
Part II–The filtrate is acidified with dil. CH_3COOH and then potassium chromate solution is added. Yellow ppt.
Pb^{2+} present.</td></tr>
<tr><td>Residue:</td><td>Filtrate:</td></tr>
<tr><td>Black.
It is dissolved in aqua regia, concentrated, cooled, diluted and $SnCl_2$ solution is added. White or grey ppt.
Hg_2^{2+} present.</td><td>It may contain $[Ag(NH_3)_2]Cl$. The filtrate is divided into two parts.
Part I—Dil. HNO_3 is added. White precipitate. Ag^+ present.
Part II—Few drops of KI solution are added. Pale yellow ppt.
Ag^+ present.</td></tr>
</table>

Things to Remember

(i) The precipitate obtained in Group I is chlorides of lead, silver and mercury(ous).

$$Pb^{2+} + 2Cl^- \rightarrow PbCl_2 \downarrow$$

$$Ag^+ + Cl^- \rightarrow AgCl \downarrow$$

$$2Hg^+ + 2Cl^- \rightarrow Hg_2Cl_2 \downarrow$$

(ii) $PbCl_2$ is soluble in hot water whereas AgCl and Hg_2Cl_2 are not soluble.

(iii) AgCl dissolves in ammonia solution whereas Hg_2Cl_2 turns black.

$$AgCl + 2NH_4OH \rightarrow \underset{\text{(Soluble)}}{\left[Ag(NH_3)_2\right]Cl} + 2H_2O$$

$$\underbrace{\underset{\text{(White ppt.)}}{Hg_2Cl_2 + 2NH_3 \rightarrow Hg(NH_2)Cl} + \underset{\text{(Black)}}{Hg}}_{\text{Black ppt.}} + NH_4Cl$$

(iv) Aqua regia converts mercuric amino chloride and mercury into mercuric chloride.

$$HNO_3 + 3HCl \rightarrow NOCl + 2H_2O + 2Cl$$

$$2Hg(NH_2)Cl + 6Cl \rightarrow 2HgCl_2 + 4HCl + N_2$$

$$Hg + 2Cl \rightarrow HgCl_2$$

Stannous chloride reduces mercuric chloride into white or grey ppt.

$$2HgCl_2 + SnCl_2 \rightarrow \underset{\text{(white)}}{Hg_2Cl_2} + SnCl_4$$

$$Hg_2Cl_2 + SnCl_2 \rightarrow \underset{\text{(Grey)}}{2Hg} + SnCl_4$$

(v) Diammine silver(I) chloride reacts with conc. HNO_3 to give white ppt. of AgCl.

$$[Ag(NH_3)_2]Cl + 2HNO_3 \rightarrow \underset{\text{(white ppt.)}}{AgCl} + 2NH_4NO_3$$

(vi) $PbCl_2$ is soluble in hot water and insoluble in cold water.

(vii) $PbCl_2$ reacts with potassium chromate solution to give yellow ppt. of lead chromate.

$$PbCl_2 + K_2CrO_4 \rightarrow \underset{\text{(yellow ppt.)}}{PbCrO_4} + 2KCl$$

Table 9: Analysis of Group I in Presence of Tungsten; W

The precipitate may contain $PbCl_2$, AgCl, Hg_2Cl_2 and tungstic acid. The ppt. is washed on the filter paper first with dil. HCl and then with small amount of cold water. The ppt. is transferred to a hard glass test tube and boiled with 25 c.c. water and filtered hot.	
Residue:	Filtrate:
It may contain Hg_2Cl_2, AgCl and tungstic acid. The ppt. is washed with hot water several times until the washings give no ppt. with K_2CrO_4 solution. About 15 c.c. warm NH_4OH (1:1) is poured over the ppt. and the filtrate collected.	It may contain $PbCl_2$. The filtrate is divided into two parts.

(Contd...)

<table>
<tr><td>Residue:</td><td colspan="2">Filtrate:</td><td rowspan="4">Part I—It is cooled. White crystals.
Pb^{2+} present.
Part II—It is treated with KI solution. Yellow ppt.
Pb^{2+} present.</td></tr>
<tr><td rowspan="3">Black.
It is dissolved in aqua regia, concentrated, cooled, diluted and $SnCl_2$ solution is added. White or grey ppt.
Hg_2^{2+} present</td><td colspan="2">It may contain $[Ag(NH_3)_2]Cl$ and ammonium tungstate. It is neutralised with dil. HCl. Dil. NH_4OH is added just to redissolve any ppt. which forms. Now KI solution is added and filtered.</td></tr>
<tr><td>Residue:</td><td>Filtrate:</td></tr>
<tr><td>Pale yellow ppt. of AgI.
Ag^+ present</td><td>It may contain tungstate. It is evaporated to a small volume. The solution is acidified with dil. HCl, and then 3 c.c. stannous chloride solution is added. It is boiled. Further 3 c.c. conc. HCl is added and boiled.
Blue ppt. or colouration.
W present</td></tr>
</table>

Table 10: Separation of Group II into IIA and IIB

<table>
<tr><td colspan="2">The precipitate may contain sulphides of Group II metals. The precipitate is transferred to a porcelain dish. It is washed with ammonium chloride solution that has been saturated with H_2S. Now 5-10 c.c. ammonium polysulphide solution is added and heated to 50-60°C for 3-4 minutes with constant stirring. It is filtered. The precipitate is washed with dilute (1:100) ammonium polysulphide solution.</td></tr>
<tr><td>Residue:</td><td>Filtrate:</td></tr>
<tr><td>May contain HgS, PbS, Bi_2S_3, CuS, CdS.
Group IIA present.</td><td>May contain solution of thio-salts of As, Sb, Sn, Mo. The filtrate is acidified with conc. HCl (test with litmus paper) and warmed gently. A coloured ppt. is obtained.
Group IIB present.</td></tr>
</table>

Things to Remember

(i) The precipitate of Group II is first washed with ammonium chloride solution that has been saturated with H_2S. Here H_2S prevents conversion of CuS to $CuSO_4$ by atmospheric oxidation.

(ii) Precipitates of Group IIA are insoluble in ammonium polysulphide whereas precipitates of IIB are soluble. Ammonium polysulphide is represented as $(NH_4)_2S_2$. Number of sulphur atoms in a molecule of ammonium polysulphide may vary from 2-5.

(iii) Precipitates of Group IIB dissolve in ammonium poly sulphide forming thio salts.

$As_2S_3 + 4S_2^{2-} \rightarrow S_3^{2-} + 2AsS_4^{3-}$ (Thioarsenate)

$Sb_2S_3 + 4S_2^{2-} \rightarrow S_3^{2-} + 2SbS_4^{3-}$ (Thioantimonate)

$Sb_2S_5 + 6S_2^{2-} \rightarrow 3S_3^{2-} + 2SbS_4^{3-}$ (Thioantimonate)

$SnS + S_2^{2-} \rightarrow SnS_3^{2-}$ (Thiostannate)

$SnS_2 + 2S_2^{2-} \rightarrow S_3^{2-} + SnS_3^{2-}$ (Thiostannate)

(iv) Upon acidification of the solution of the thiosalts, sulphides of Group IIB metals are precipitated.

$2AsS_4^{3-} + 6H^+ \rightarrow As_2S_5 \downarrow + 3H_2S \uparrow$

$2SbS_4^{3-} + 6H^+ \rightarrow Sb_2S_5 \downarrow + 3H_2S \uparrow$

$SnS_3^{2-} + 2H^+ \rightarrow SnS_2 \downarrow + H_2S \uparrow$

(v) On acidification, sulphur is also precipitated which is a disadvantage of this method.

$S_2^{2-} + 2H^+ \rightarrow S \downarrow + H_2S$

Sulphur is precipitated due to presence of excess reagent.

(vi) If the precipitate is completely soluble in ammonium polysulphide solution Group IIA is absent.

Table 11: Analysis of Group IIA

The ppt. may contain HgS, PbS, Bi_2S_3, CdS, and CuS. The ppt. is transferred to a porcelain basin. About 5 c.c. dil. HNO_3 is added, boiled gently for few minutes, filtered and washed with a little water.	
Residue:	Filtrate:
Black. It is dissolved in	May contain Pb^{2+}, Bi^{3+}, Cu^{2+} and Cd^{2+}. Excess concentrated ammonia solution is added until precipitation is complete and filtered.

(Contd...)

<table>
<tr>
<td rowspan="6">aqua regia, concentrated, cooled, diluted and then $SnCl_2$ solution is added. White or grey ppt.
Hg^{2+} present.</td>
<td colspan="2">Residue:</td>
<td colspan="2">Filtrate:</td>
</tr>
<tr>
<td colspan="2">May contain $Bi(OH)_3$ and $Pb(OH)_2$. About 5 c.c. of NaOH solution is added and warmed. Then it is filtered.</td>
<td colspan="2" rowspan="3">May contain $[Cu(NH_3)_4]^{2+}$ and $[Cd(NH_3)_4]^{2+}$.
Case I: If colourless, Cu is absent. Then only Cd is tested by passing H_2S. Yellow ppt.
Cd^{2+} present.
Case II: If blue, solution is divided into two unequal parts.
Smaller Part: It is acidified with dil. CH_3COOH and potassium ferrocyanide is added. Reddish-brown ppt.
Cu^{2+} present.
Larger Part: KCN solution is added dropwise until it is decolourised. Now H_2S is passed. Yellow ppt. of Cds.
Cd^{2+} present.
In the case II an alternative procedure to avoid the use of highly poisonous KCN is to acidify the solution with 6(N)HCl, saturate with solid NH_4Cl and pass H_2S followed by filtration.</td>
</tr>
<tr>
<td>Residue:</td>
<td>Filtrate:</td>
</tr>
<tr>
<td>White.
May contain $Bi(OH)_3$. It is washed with little water. Sodium tetrahydroxostannate(II) solution is poured over filter paper. Blackening of ppt.
Bi^{3+} present.</td>
<td>May contain $[Pb(OH)_4]^{2-}$. It is acidified with dil. CH_3COOH and then K_2CrO_4 solution is added. Yellow ppt.
Pb^{2+} present.</td>
</tr>
<tr>
<td colspan="2" rowspan="2"></td>
<td>Residue:</td>
<td>Filtrate:</td>
</tr>
<tr>
<td>Black precipitate of Cu^{2+} present.</td>
<td>May contain Cd. Sodium acetate is added and then H_2S is passed. Yellow precipitate.
Cd^{++} present.</td>
</tr>
</table>

Things to Remember

(i) HgS is insoluble in dil. HNO_3, while PbS, Bi_2S_3, CuS, and CdS all dissolve.

$$3PbS + 8H^+ + 2NO_3^- \rightarrow 3Pb^{2+} + 3S\downarrow + 2NO\uparrow + 4H_2O$$

$$Bi_2S_3 + 8H^+ + 2NO_3^- \rightarrow 2Bi^{3+} + 3S\downarrow + 2NO\uparrow + 4H_2O$$

$$3CuS + 8H^+ + 2NO_3^- \rightarrow 3Cu^{2+} + 3S\downarrow + 2NO\uparrow + 4H_2O$$

$$3CdS + 8H^+ + 2NO_3^- \rightarrow 3Cd^{2+} + 3S\downarrow + 2NO\uparrow + 4H_2O$$

(ii) Black residue of HgS dissolves in aqua regia forming mercuric chloride.

$$3HgS + 2HNO_3 + 6HCl \rightarrow 3HgCl_2 + 3S + 2NO + 4H_2O$$

Stannous chloride reduces mercuric chloride into white ppt. of mercurous chloride.

$$2HgCl_2 + SnCl_2 \rightarrow Hg_2Cl_2 + SnCl_4$$

Mercurous chloride may further be reduced to mercury.

$$Hg_2Cl_2 + SnCl_2 \rightarrow \underset{\text{(Black)}}{2Hg} + SnCl_4$$

(iii) Cu^{2+} and Cd^{2+} react with excess of ammonia to form complexes and remain in solution.

$$Cu^{2+} + 4NH_3 \rightarrow \left[Cu(NH_3)_4\right]^{2+}$$

$$Cd^{2+} + 4NH_3 \rightarrow \left[Cd(NH_3)_4\right]^{2+}$$

Bi^{3+} and Pb^{2+} however get precipitated as hydroxides.

$$Bi^{3+} + 3NH_3 + 3H_2O \rightarrow Bi(OH)_3\downarrow + 3NH_4^+$$

$$Pb^{2+} + 2NH_3 + 2H_2O \rightarrow Pb(OH)_2\downarrow + 2NH_4^+$$

(iv) $Pb(OH)_2$ dissolves in NaOH, while $Bi(OH)_3$ remains unaffected.

$$Pb(OH)_2 + 2OH^- \rightarrow [Pb(OH)_4]^{2-}$$

(v) Sodium tetrahydroxostannate(II) solution is prepared by adding NaOH to $SnCl_2$ solution until the white ppt. of $Sn(OH)_2$ dissolves completely. The test of bismuth is based on the formation of bismuth metal.

$$2Bi^{3+} + 3[Sn(OH)_4]^{2-} + 6OH^- \rightarrow 2Bi\downarrow + 3[Sn(OH)_6]^{2-}$$

(vi) $[Pb(OH)_4]^{2-}$ reacts with acetic acid to form lead acetate.

$$[Pb(OH)_4]^{2-} + 4CH_3COOH \rightarrow (CH_3COO)_2Pb + 2CH_3COO^- + 4H_2O$$

Lead acetate reacts with K_2CrO_4 solution to give yellow ppt. of lead chromate.

$$(CH_3COO)_2Pb + K_2CrO_4 \rightarrow \underset{(Yellow)}{PbCrO_4} \downarrow + 2CH_3COOK$$

(vii) $[Cd(NH_3)_4]^{2+}$ reacts with H_2S to form yellow ppt. of CdS.

$$[Cd(NH_3)_4]^{2+} + H_2S \rightarrow \underset{(Yellow)}{CdS} \downarrow + 2NH_4^+ + 2NH_3$$

This test of cadmium is done when copper is absent.

(viii) Acetic acid decomposes the blue $[Cu(NH_3)_4]^{2+}$ complex to Cu^{2+}, which in turn is precipitated by $K_4[Fe(CN)_6]$.

$$[Cu(NH_3)_4]^{2+} + 4CH_3COOH \rightarrow Cu^{2+} + 4NH_4^+ + 4CH_3COO^-$$

$$2Cu^{2+} + [Fe(CN)_6]^{4-} \rightarrow Cu_2[Fe(CN)_6] \downarrow$$

(ix) If copper is present, then to test cadmium, KCN solution is added.

$$[Cd(NH_3)_4]^{2+} + 4CN^- \rightarrow [Cd(CN)_4]^{2-} + 4NH_3$$

$$2[Cu(NH_3)_4]^{2+} + 10CN^- \rightarrow 2[Cu^I(CN)_4]^{3-} + (CN)_2 \uparrow + 8NH_3$$

The tetracyanocuprate(I) complex is so stable that H_2S does not produce a ppt. From the tetracyanocadmiate(II) complex yellow CdS is precipitated with H_2S.

$$[Cd(CN)_4]^{2-} + H_2S + 2NH_3 \rightarrow \underset{(Yellow)}{CdS} \downarrow + 2NH_4^+ + 4CN^-$$

(x) The filtrate curtaining Cu^{2+} and Cd^{2+} is made strongly acidic with 6(N) HCl and H_2S is passed whereby Cu^{2+} is precipitated as black CuS. CdS is not precipitated from strongly acidic solutions. After the removal of CuS, the solution is treated with sodium acetate to reduce the acid content and H_2S gas is passed through this solution whereby Cd^{2+} is precipitated as yellow CdS.

Table 12: Analysis of Group IIB

The ppt. obtained by the treatment of ammonium polysulphide followed by conc. HCl may contain sulphides of As, Sb, Sn along with light yellow or white sulphur. It is filtered and washed. The ppt. is transferred to a conical flask. About 15 c.c. conc. HCl is added, boiled gently for 5 minutes with the funnel in the mouth of the flask. It is diluted with about 5 c.c. water and filtered.	
Residue:	Filtrate:
Yellow. May contain As_2S_5, As_2S_3, and S. The ppt. is dissolved in 3-4 c.c. warm ammonia and filtered if necessary. About 3-4 c.c. 3% H_2O_2 solution is added and warmed for a few minutes to oxidise them into arsenate. Few c.c. of $Mg(NO_3)_2$ reagent is added, stirred and allowed to stand. White ppt. As^{3+} present. Now 1 c.c. of $AgNO_3$ containing CH_3COOH is added on the ppt. Brownish red residue. As^{3+} present.	May contain Sb^{3+} and Sn^{4+}. It is boiled to expel H_2S, cooled and the cold solution divided in three parts. Part I: About 1-2 g iron powder is added and warmed until a steady evolution of hydrogen starts. The reduction is allowed to proceed for 3-4 minutes, then filtered. The filtrate is added to $HgCl_2$ solution. White or grey ppt. Sn^{4+} present. Part II: (If Sn is present) It is made alkaline with ammonia solution. 1-2 g solid oxalic acid is added, boiled, and H_2S is passed for 1 minute into hot filtrate. Orange ppt. Sb^{3+} present. The filtrate obtained after the removal of the above precipitate is made just alkaline with NH_4OH solution and then acidified with dilute acetic acid. The solution is heated to boiling and H_2S is passed. A brownish yellow ppt. Sn^{4+} present. Part III: (If Sn is absent) It is diluted with water, then H_2S is passed. Orange ppt. Sb^{3+} present.

Things to Remember

(i) When the sulphides of Group IIB metals are treated with conc. HCl, Sb_2S_5 and SnS_2 dissolve in it, while As_2S_5 and As_2S_3 remain unaffected.

$$Sb_2S_5 + 6HCl \rightarrow 2Sb^{3+} + 3H_2S + 2S \downarrow + 6Cl^-$$

$$SnS_2 + 4HCl \rightarrow Sn^{4+} + 2H_2S + 4Cl^-$$

(ii) The As_2S_5 and As_2S_3 precipitates are converted by NH_3 and H_2O_2 to soluble arsenate.

$$As_2S_5 + 16NH_3 + 20H_2O_2 \rightarrow 2AsO_4^{3-} + 16NH_4^+ + 5SO_4^{2-} + 12H_2O$$

$$As_2S_3 + 12NH_3 + 14H_2O_2 \rightarrow 2AsO_4^{3-} + 12NH_4^+ + 3SO_4^{2-} + 8H_2O$$

(iii) The identification reactions of As are as follows:

$$AsO_4^{3-} + NH_4^+ + Mg^{2+} \rightarrow \underset{\text{(White)}}{MgNH_4AsO_4} \downarrow$$

$$AsO_4^{3-} + 3Ag^+ \rightarrow \underset{\text{(Brown)}}{Ag_3AsO_4} \downarrow$$

(iv) Iron reduces Sn^{4+} and Sb^{3+} as follows:

$$Sn^{4+} + Fe \rightarrow Sn^{2+} + Fe^{2+}$$

$$2Sb^{3+} + 3Fe \rightarrow 2Sb \downarrow + 3Fe^{2+}$$

Now Sn^{2+} reduces $HgCl_2$ into white ppt. of Hg_2Cl_2.

$$Sn^{2+} + 2HgCl_2 \rightarrow Hg_2Cl_2 + Sn^{4+} + 2Cl^-$$

If Sn^{2+} is present in excess, grey mercury is formed.

$$Sn^{2+} + Hg_2Cl_2 \rightarrow \underset{\text{(Grey)}}{2Hg} \downarrow + Sn^{4+} + 2Cl^-$$

(v) The filtrate from the HCl treatment may contain Sb^{3+} and Sn^{4+} ions. When oxalic acid is added, Sn^{4+} forms stable trioxalatostannate(IV) complex from which H_2S cannot produce a precipitate. Sb_2S_3 is precipitated under these conditions.

$$Sn^{4+} + 3(COO)_2^{2-} \rightarrow \underset{\text{(Stable)}}{\left[Sn\{(COO)_2\}_3\right]^{2-}}$$

$$2Sb^{3+} + 3H_2S + 6NH_3 \rightarrow \underset{\text{(Orange ppt.)}}{Sb_2S_3 \downarrow} + 6NH_4^+$$

(vi) If the filtrate from the HCl treatment contains only Sb^{3+}, H_2S is directly passed to it.

$$2Sb^{3+} + 3H_2S \rightarrow \underset{\text{(Orange ppt.)}}{Sb_2S_3 \downarrow} + 6H^+$$

Table 13: Analysis of Group IIB in Presence of Molybdenum; Mo

<table>
<tr><td colspan="3">The ppt. obtained by the treatment of ammonium polysulphide followed by conc. HCl may contain sulphides of As, Sb, Sn, Mo along with sulphur. The ppt. is boiled gently with conc. HCl for 5 minutes with funnel in mouth of flask. It is diluted with 2-3 c.c. water, and filtered.</td></tr>
<tr><td colspan="2">Residue:</td><td>Filtrate:</td></tr>
<tr><td colspan="2">May contain As_2S_3, As_2S_5, and MoS_3. The residue is washed with dil. HCl followed by water. It is now treated with 3-4 c.c. of 4 N - NH_4OH solution, warmed for 3 minutes and filtered.</td><td rowspan="3">May contain Sb^{3+} and Sn^{4+}. It is boiled to expel H_2S and the cold solution divided in three parts.

Part I: About 1-2 g iron powder is added and warmed until a steady evolution of hydrogen starts. The reduction is allowed to proceed for 3-4 minutes, then filtered. The filtrate is added to $HgCl_2$ solution.

White or grey ppt.

Sn^{4+} present.

Part II: (If Sn is present) It is made alkaline with ammonia solution. 1-2 g solid oxalic acid is added, boiled, and H_2S is passed for 1 minute into hot filtrate.

Orange ppt.

Sb^{3+} present.

Part III: (If Sn is absent) It is diluted with water, then H_2S is passed.

Orange ppt.

Sb^{3+} present.</td></tr>
<tr><td>Residue:</td><td>Filtrate:</td></tr>
<tr><td>May contain undissolved MoS_3. It is dissolved in conc. HCl and Br_2 water. It is boiled to expel Br_2. About 1-2 c.c. $SnCl_2$ solution is then added. Blue colouration. Ammonium thiocyanate solution is added.

Red colouration.

Mo present.</td><td>About 3-4 c.c. H_2O_2 is added and warmed for a few minutes to oxidise into arsenate. Further a few c.c. of $Mg(NO_3)_2$ reagent is added, stirred and allowed to stand.

White ppt.

As^{3+} present.</td></tr>
</table>

Table 14: Analysis of Group III

The ppt. may contain $Fe(OH)_3$, $Cr(OH)_3$, $Al(OH)_3$, and a little $MnO_2 \cdot xH_2O$. It is washed with a little hot 1 per cent NH_4Cl solution. The ppt. is transferred with the aid of 5-10 c.c. water to a small evaporating basin. About 1 g of sodium peroxide (or 10 c.c. of NaOH solution and 10 c.c. of 10 volume H_2O_2) is added, gently boiled for 2-3 minutes and filtered.	
Residue:	Filtrate:
May contain $Fe(OH)_3$ and $MnO_2 \cdot xH_2O$. The residue is washed with a little hot water. It is divided into two parts. Part I: It is dissolved in 1 ml. dil. HNO_3 (1:1) with the aid (if necessary) of 3-4 drops of 3% H_2O_2 solution or 1 drop of saturated H_2SO_3 solution. The solution is boiled, cooled and about 0.1 g sodium bismuthate is added with shaking. It is allowed to stand. Violet colouration. Mn^{2+} present. Part II: Another part of residue is dissolved in warm dil. HCl (filter if necessary). Potassium ferrocyanide solution is added to it. Blue ppt. Fe^{3+} present. Or, another part of residue is dissolved in dil. HCl and few drops of potassium thiocyanate is added. Deep red colouration. Fe^{3+} present.	May contain CrO_4^{2-} (yellow) and $[Al(OH)_4]^-$ (colourless). The filtrate is divided into two parts. Part I: First part of the filtrate is acidified with acetic acid and then lead acetate solution is added. Yellow ppt. Cr^{3+} present. Part II: Second part of the filtrate is acidified with dil. HCl and then dil. ammonia solution is added to make it just alkaline. It is heated to boiling and filtered. White gelatinous ppt. Al^{3+} present.

Things to Remember

(i) The precipitate of Group III may contain Mn as $MnO_2 \cdot xH_2O$, although Mn has been placed in Group IV. It is due to oxidation of $Mn(OH)_2$ by air or conc. HNO_3 (used to oxidise Fe^{2+} into Fe^{3+}).

$$\underset{\text{(Soluble)}}{Mn(OH)_2} + \underset{\text{(air)}}{\frac{1}{2}O_2} \rightarrow MnO_2 + H_2O$$

$$MnO_2 + xH_2O \rightarrow \underset{\text{(ppt.)}}{MnO_2 \cdot xH_2O}$$

(ii) When the ppt. of Group III is boiled with NaOH and H_2O_2, aluminium hydroxide and chromium hydroxide get dissolved, while ferric hydroxide and $MnO_2{\cdot}xH_2O$ remain unaffected.

$$Al(OH)_3 + OH^- \rightarrow \underset{\text{(Soluble)}}{[Al(OH)_4]^-}$$

$$2Cr(OH)_3 + 3H_2O_2 + 4OH^- \rightarrow \underset{\text{(Soluble)}}{2CrO_4^{2-}} + 8H_2O$$

The excess H_2O_2 is decomposed upon boiling to avoid any reduction of chromate upon acidification.

(iii) In the test of chromium, yellow ppt. is that of lead chromate.

$$CrO_4^{2-} + (CH_3COO)_2Pb \rightarrow \underset{\text{(Yellow ppt.)}}{PbCrO_4} + 2CH_3COO^-$$

(iv) In the test of aluminium, white gelatinous ppt. is that of aluminium hydroxide.

$$[Al(OH)_4]^- + NH_4^+ \rightarrow Al(OH)_3 \downarrow + NH_3 + H_2O$$

(v) $MnO_2{\cdot}xH_2O$ dissolves in dil. HNO_3 and a few drops of H_2O_2 solution.

$$MnO_2 + H_2O_2 + 2H^+ \rightarrow Mn^{2+} + O_2 + 2H_2O$$

Mn^{2+} ion is oxidised to MnO_4^- with sodium bismuthate, $NaBiO_3$.

$$2Mn^{2+} + 5NaBiO_3 + 14H^+ \rightarrow 2MnO_4^- + 5Bi^{3+} + 5Na^+ + 7H_2O$$

(vi) $Fe(OH)_3$ dissolves in dil. HCl. Potassium ferrocyanide gives blue ppt. and potassium thiocyanate gives blood red colouration with $Fe(OH)_3$.

$$4Fe^{3+} + 3[Fe^{II}(CN)_6]^{4-} \rightarrow \underset{\text{(Prussian blue)}}{Fe_4^{III}[Fe^{II}(CN)_6]_3}$$

$$Fe^{3+} + 6SCN^- \rightarrow \underset{\text{(Blood red)}}{[Fe^{III}(SCN)_6]^{3-}}$$

Table 15: Analysis of Group III in Presence of Ce and V

<table>
<tr><td colspan="3">The ppt. of Group III is dissolved in minimum volume of dil. HCl. This weakly acid solution is poured into an equal volume of 30% H_2O_2 and NaOH solution. It is boiled for 5 minutes, filtered and the ppt. washed with hot 2 per cent NH_4NO_2.</td></tr>
<tr><td>Residue:</td><td colspan="2">Filtrate:</td></tr>
<tr><td rowspan="3">May contain $Fe(OH)_3$ and $CeO_3^{2-}\cdot xH_2O$, and some $MnO_2\cdot xH_2O$. The residue is dissolved in dil. HCl and the solution is divided into three parts.

Part I: Potassium thiocyanate solution is added.

Blood red colouration.

Fe^{3+} present.

Part II: Excess of saturated oxalic acid solution is added.

White ppt.

Ce^{4+} present.

Part III: It is evaporated with conc. H_2SO_4 to expel HCl, cooled, diluted, and then dil. HNO_3 is added. At last sodium bismuthate is added with stirring and allowed to stand.

Purple colouration.

Mn^{2+} present.</td><td colspan="2">May contain CrO_4^{2-}, $[Al(OH)_4]^-$, and VO_3^-. The precipitate is acidified with dil. HNO_3 and about 3-4 c.c. 0.25 M-$Pb(NO_3)_2$ solution is added. Then about 2 g ammonium acetate is added, stirred well, filtered and washed with hot water.</td></tr>
<tr><td>Residue:</td><td>Filtrate:</td></tr>
<tr><td>May contain $PbCrO_4$ and $Pb(VO_3)_2$. The residue is dissolved in minimum vol. of hot 2M-HNO_3 (5-6 c.c.), cooled, and transferred to a small separating funnel. An equal volume of amyl alcohol, and a little 6 per cent H_2O_2 are added. It is shaken well and the two layers allowed to separate.

Blue colouration in the upper layer.

Cr^{3+} present.

Red to brownish-red colouration in the lower layer.

V^{3+} present.</td><td>May contain Al^{3+}, and excess of Pb^{2+}. H_2S is passed through the filtrate to remove lead. PbS is filtered off. The filtrate is boiled to expel H_2S. Now it is neutralised with NH_3 solution, cooled and poured into an excess of conc. $(NH_4)_2CO_3$ solution.

White gelatinous ppt.

Al^{3+} present.</td></tr>
</table>

Table 16: Analysis of Group IV

The ppt. may contain CoS, NiS, MnS and ZnS. It is washed with 1 per cent NH_4Cl solution to which 1 per cent by volume of $(NH_4)_2S$ has been added. The ppt. is transferred to a small beaker. About 5 c.c. water and 5 c.c. 2M-HCl are added, stirred well, allowed to stand for 2-3 minutes and filtered.

(*Contd...*)

<table>
<tr><td>Residue:</td><td colspan="2">Filtrate:</td></tr>
<tr><td rowspan="4">The residue is dissolved in a mixture of 1.5 ml M-NaOCl solution and 0.5 c.c. dil. HCl. Further 1 c.c. dil. HCl is added and boiled until all Cl_2 is expelled. It is cooled and diluted to about 4 c.c. The solution is divided into two equal parts.
Part I: About 1 c.c. amyl alcohol, and 2 g solid NH_4SCN are added to the first part. It is shaken well.
Blue colouration in amyl alcohol layer.
Co^{2+} present.
Part II: About 2 c.c. NH_4Cl solution is added followed by NH_3 solution until alkaline. Then dimethyl glyoxime reagent is added in excess.
Red ppt.
Ni^{2+} present.</td><td colspan="2">May contain Mn^{2+} and Zn^{2+}. The filtrate is boiled to expel H_2S. It is cooled, and excess NaOH solution added followed by 1 c.c. 3% H_2O_2 solution. The solution is boiled for 3 minutes and filtered.</td></tr>
<tr><td>Residue:</td><td>Filtrate:</td></tr>
<tr><td>$MnO_2 \cdot xH_2O$. The residue is dissolved in 5 c.c. dil HNO_3 (1:1) with the addition of a few drops of 3% H_2O_2 solution, if necessary. It is boiled to decompose excess H_2O_2 and cooled. Now 0.05 g $NaBiO_3$ is added, stirred and allowed to settle.
Purple colouration.
Mn^{2+} present.</td><td>May contain $[Zn(OH)_4]^{2-}$. The filtrate is acidified with acetic acid and then H_2S is passed.
White ppt.
Zn^{2+} present.</td></tr>
</table>

Things to Remember

(i) MnS and ZnS dissolve readily in cold and very dilute HCl, while NiS and CoS remain almost undissolved.

$$MnS + 2H^+ \rightarrow Mn^{2+} + H_2S \uparrow$$

$$ZnS + 2H^+ \rightarrow Zn^{2+} + H_2S \uparrow$$

(ii) Mixture of CoS and NiS is dissolved either in aqua regia or in NaOCl and HCl.

$$3CoS + 2HNO_3 + 6HCl \rightarrow 3Co^{2+} + 2NO \uparrow + 3S \downarrow + 6Cl^- + 4H_2O$$

$$CoS + OCl^- + 2H^+ \rightarrow Co^{2+} + S \downarrow + Cl^- + H_2O$$

Co^{2+} ion give blue colouration with NH_4SCN.

$$Co^{2+} + 4SCN^- \rightarrow \underset{\text{(Blue)}}{[Co(SCN)_4]^{2-}}$$

Test of Ni^{2+} can be done with dimethyl glyoxime reagent in presence of Co^{2+} provided excess reagent is added.

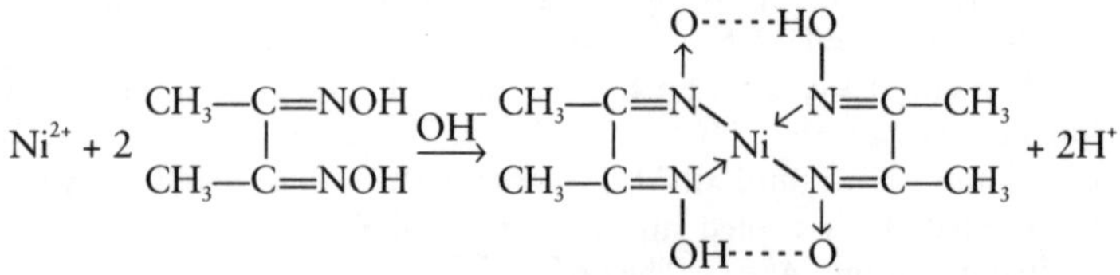

(Nickel dimethyl glyoximate; red)

(iii) The solution containing Mn^{2+} and Zn^{2+} and possibly traces of Co^{2+} and Ni^{2+} is boiled to remove H_2S and then warmed with little H_2O_2. H_2O_2 oxidises Co^{2+} to Co^{3+} and Mn^{2+} to Mn^{4+}. $Co(OH)_2$ is slightly soluble but $Co(OH)_3$ is insoluble in excess NaOH solution.

$$Co^{2+} \xrightarrow{H_2O_2} Co^{3+}$$

$$Mn^{2+} \xrightarrow{H_2O_2} Mn^{4+} \left(MnO_2 \cdot xH_2O\right)$$

$Zn(OH)_2$ will dissolve in excess of NaOH solution forming $[Zn(OH)_4]^{2-}$ while $MnO_2 \cdot xH_2O$, and traces of $Co(OH)_3$ and $Ni(OH)_2$ remain insoluble.

Zn may be tested by passing H_2S.

$$\left[Zn(OH)_4\right]^{2-} + H_2S \rightarrow \underset{(White)}{ZnS} \downarrow + 2OH^- + 2H_2O$$

Mn is readily identified by dissolving $MnO_2 \cdot xH_2O$ in dil. HNO_3 and a little H_2O_2 and adding sodium bismuthate.

$$MnO_2 + H_2O_2 + 2H^+ \rightarrow Mn^{2+} + O_2 + 2H_2O$$

$$2Mn^{2+} + 5NaBiO_3 + 14H^+ \rightarrow \underset{(Purple)}{2MnO_4^-} + 5Bi^{3+} + 5Na^+ + 7H_2O$$

Table 17: Analysis of Group V

The ppt. may contain $BaCO_3$, $SrCO_3$, and $CaCO_3$. It is washed and then dissolved in about 20 c.c. hot dil. CH_3COOH. To test the presence of Ba, about 2 c.c. of above solution is taken, boiled and potassium chromate solution is added. Yellow ppt. indicates the presence of Ba.

(Contd...)

<table>
<tr><td colspan="2">Ba Present:
The remaining solution is heated to boiling and a slight excess of potassium chromate solution is added until the solution assumes an orange tint. The ppt. of $BaCrO_4$ is filtered off. The filtrate is treated with a few drops of liquor ammonia and excess of ammonium carbonate solution. It is warmed. A white ppt. indicates the presence of Sr and/or Ca. The ppt. is washed and dissolved in about 10 c.c. hot dil. CH_3COOH. It is boiled to remove excess CO_2. Call it solution (A).
Ba Absent:
The remaining solution is subjected to test Sr and Ca. Call it solution (B).
Now solution A (if Ba present) or solution B (if Ba absent) is taken and excess of ammonium sulphate solution is added. It is heated to boiling and allowed to stand for 5 minutes. Now it is filtered and washed.</td></tr>
<tr><td>Residue:</td><td>Filtrate</td></tr>
<tr><td>White ppt.
Sr^{2+} present.</td><td>The filtrate is concentrated and a little ammonia solution is added. Then ammonium oxalate solution is added, boiled and allowed to stand.
White ppt.
Ca^{2+} present.</td></tr>
</table>

Things to Remember

(i) The precipitate of Group V dissolves in dil. acetic acid.

$$BaCO_3 + 2CH_3COOH \rightarrow (CH_3COO)_2\,Ba + H_2O + CO_2 \uparrow$$

$$SrCO_3 + 2CH_3COOH \rightarrow (CH_3COO)_2\,Sr + H_2O + CO_2 \uparrow$$

$$CaCO_3 + 2CH_3COOH \rightarrow (CH_3COO)_2\,Ca + H_2O + CO_2 \uparrow$$

(ii) Ba is identified by the formation of yellow ppt. of $BaCrO_4$.

$$Ba^{2+} + CrO_4^{2-} \rightarrow \underset{\text{(Yellow)}}{BaCrO_4} \downarrow$$

Calcium and strontium chromates are soluble in acetic acid and so they do not get precipitated.

(iii) The addition of $(NH_4)_2SO_4$ solution to a solution of Sr and Ca acetates produces the ppt. of $SrSO_4$ only; calcium forms a soluble salt.

$$Sr^{2+} + SO_4^{2-} \rightarrow \underset{(White)}{SrSO_4} \downarrow$$

(iv) The calcium in the filtrate is precipitated as oxalate by the addition of ammonium oxalate solution.

$$Ca^{2+} + C_2O_4^{2-} \rightarrow \underset{(White)}{CaC_2O_4} \downarrow$$

Table 18: Analysis of Group VI

<table>
<tr><td colspan="3">The filtrate from Group V is concentrated to half its bulk. To test the presence of Mg, a portion (5 c.c.) of the concentrated solution is taken. A pinch of solid ammonium chloride, a few drops of liquor ammonia and excess of ammonium hydrogen phosphate solution are added. It is shaken vigorously and the walls of the test tube scratched. If a white ppt. forms, it is diluted with an equal volume of water. If the ppt. remains undissolved, the presence of Mg is confirmed.
If Mg is present, whole of the concentrated filtrate is treated in the above manner; filtered.
If Mg is absent, tests for Na and K are done directly from the concentrated filtrate.</td></tr>
<tr><td>Residue:</td><td colspan="2">Filtrate:</td></tr>
<tr><td rowspan="3">White.
Mg^{2+}
present.</td><td colspan="2">This filtrate (in case Mg present) or concentrated filtrate from Group V (in case of absence of Mg) may contain Na and/or K. The filtrate is evaporated in either case in a basin. The residue is further heated until no more fumes of ammonium salts (previously added) are evolved. If no residue is left Na and K are absent. If a solid residue is obtained, Na and/or K are indicated.
A little of the residue is kept for flame test and the remaining mass is dissolved in about 10 c.c. water. The solution is divided into two parts.</td></tr>
<tr><td>For Sodium:</td><td>For Potassium:</td></tr>
<tr><td>Residue: Flame test is done.
Golden yellow flame.
Na^+ present.
Solution: Potassium dihydrogen antimonate is added and the inner wall of the test tube rubbed.
White ppt.
Na^+ present.</td><td>Residue: Flame test is done. Violet flame appearing violet red through blue glass.
K^+ present.
Solution: Few drops of dil. CH_3COOH and equal volume of freshly prepared sodium cobaltinitrite are added and warmed.
Yellow ppt.
K^+ present.</td></tr>
</table>

Test for NH_4^+:

Original salt mixture is heated with dil. NaOH solution. NH_3 gas is evolved. It is tested by a rod moistened with conc. HCl which will give white dense fumes. NH_4^+ present.

Things to Remember

(i) The addition of NH_4Cl, NH_4OH and ammonium hydrogen phosphate into the solution of Mg^{2+} produces white crystalline ppt.

$$Mg^{2+} + NH_4OH + (NH_4)_2 HPO_4 \rightarrow \underset{\text{(White)}}{Mg(NH_4)PO_4} + 2NH_4^+ + H_2O$$

(ii) Mg interferes with the test of Na and K, so it must be completely removed.

(iii) Most of the precipitation reactions for K^+ ions are also given by NH_4^+ ions, hence the latter must be removed by evaporation to dryness and ignition.

(iv) The addition of sodium cobaltinitrite solution to solution containing K^+ ions results in precipitation of yellow ppt.

$$Na_3\left[Co^{III}(NO_2)_6\right] + 3K^+ \rightarrow \underset{\text{(Yellow ppt.)}}{K_3\left[Co(NO_2)_6\right]} + 3Na^+$$

(v) The addition of potassium dihyrogen antimonate solution to fairly strong solution containing Na^+ ions results in the precipitation of white crystalline sodium hydrogen antimonate, NaH_2SbO_4.

$$KH_2SbO_4 + Na^+ \rightarrow \underset{\text{(White ppt.)}}{NaH_2SbO_4} + K^+$$

(vi) The 'flame test' is the only reliable test for sodium and potassium. Therefore, other confirmatory tests should be performed only if the flame test is affirmative.

(vii) NH_4^+ ion has been included in Group VI, but it is tested separately by applying the test on solid salt mixture directly.

8. TREATMENT OF INORGANIC SUBSTANCES INSOLUBLE IN ACID

Substances that are not dissolved by solvents recommended before (water, HCl or HNO_3) require special methods for bringing them into solution, and are termed insolubles. They may be divided into two groups though the first group, which dissolves in aqua regia is often not included among insolubles. In that case, treatment with aqua regia constitutes one of the regular methods to prepare solution for examination of basic radicals. In this book, as in certain others, the aqua regia group is included among insolubles because it consists of a very small number of substances.

8.1. First Group

The first group of insolubles includes the sulphides: HgS, As_2S_3, Sb_2S_3, SnS_2, CoS and NiS. They are most readily brought into solution by boiling them with aqua regia—a mixture of concentrated hydrochloric acid (3 vol.) and concentrated nitric acid (1 vol.)

The insoluble is first dissolved in aqua regia. It is heated to dryness. In this stage metals of insolubles are converted into corresponding chlorides, while the acid radical (sulphide) is oxidised to elementary sulphur (globules of plastic sulphur separate out during dissolution), which may further be oxidised in part, to sulphuric acid. The residue is dissolved in water and filtered, if necessary. The solution is subjected to examine the basic radicals as before. Acid radical can be detected from sodium carbonate extract as discussed before.

8.2. Second Group

The second group of insolubles include the following substances:

(i) Halides of silver – AgCl, AgBr, AgI.

(ii) Sulphates of barium, strontium and lead — $BaSO_4$, $SrSO_4$, $PbSO_4$.

(iii) Ignited oxides of iron, chromium and tin — Fe_2O_3, Cr_2O_3, SnO_2.

(iv) Silicates.

The given salt mixture is treated with hydrochloric acid or nitric acid. The whole mixture may remain undissolved or a part of it remain undissolved. It is filtered. If a part of the mixture has dissolved, the filtrate is used to analyse the basic radical by general scheme. The residue (insoluble) is now treated to bring it in solution. The method of preparation of solution depends upon nature of insolubles. The following preliminary tests are first carried out with the insoluble portion of the mixture.

Table 19: Preliminary Tests with Insolubles

Experiment	Observation	Inference
(i) Flame test	Green Red	$BaSO_4$ $SrSO_4$
(ii) Borax bead test	Green Greenish yellow	Cr_2O_3 Fe_2O_3
(iii) Charcoal reducing	Malleable bead	AgCl, AgBr, AgI, $PbSO_4$, SnO_2
(iv) Microcosmic salt bead test	Floating skeleton	Silicate

After the idea of insoluble obtained by preliminary tests, it is treated as follows:

(i) Halides of Silver: The insoluble is thoroughly mixed with 4 times its quantity of dry fusion mixture and fused in a nickel crucible by heating on the blow pipe for 10 minutes until a tranquil melt is obtained.

$$2AgX + Na_2CO_3 \rightarrow 2Ag + 2\ NaX + CO_2 + \frac{1}{2}O_2$$

The melt after cooling is dissolved in hot water and filtered. The filtrate is used to test the acid radicals. The residue is dissolved in hot dil. HNO_3 and tested for basic radicals like silver.

(ii) Sulphates: The insoluble is fused with fusion mixture as discribed above using six times the quantity of dry fusion mixture.

$$Na_2CO_3 + BaSO_4 \rightarrow BaCO_3 + Na_2SO_4$$

The fused mass after cooling is dissolved in hot water, filtered, and washed with hot water till the filtrate no longer gives any precipitate with $BaCl_2$ solution. The filtrate is tested for acid radical and the residue for the basic radical dissolving in dil. HCl.

(iii) **Ignited Oxides:** The insoluble is fused with potassium bisulphate in a Bunsen flame. The melt after cooling is dissolved in dil. H_2SO_4. The oxides are thus changed into sulphates. The solution is used to examine the basic radicals.

(iv) **Silicates:** The insoluble is fused with fusion mixture and proceeded as (ii).

2 SEPARATION AND IDENTIFICATION OF CATIONS/ANIONS USING PAPER CHROMATOGRAPHY

1. CHROMATOGRAPHY

Chromatography is the most modern and versatile method used to separate mixtures into its components, purify compounds and also to identify them. This method was first discovered by TsWett, a Russian botanist, in 1906. The name chromatography was originally derived from the Greek word 'chroma' meaning colour and 'graphy' meaning style of drawing or writing because the method was first used for the separation of coloured substances (plant pigments) into individual components. Now this method is widely used for separation, purification and characterisation of the components of a mixture, whether coloured or colourless.

Chromatography is essentially a physical method of separation. It is defined as follows:

"The techinque of separating the components of a mixture in which separation is achieved by the differential movements of individual components through a stationary phase under the influence of a mobile phase."

In this technique, the mixture of the components is applied on either a solid or a liquid phase called 'stationary phase' as it is kept stationary. Another phase, liquid or a gas, known as the 'mobile phase' is then allowed to move slowly over the stationary phase. The mobile phase makes the components of the mixture to move along with it and selective retardation in their rates of movement separates them. The basis of retardation or separation in chromatography process is the process of adsorption or partition or both depending upon the nature of the stationary phase.

2. TYPES OF CHROMATOGRAPHY

There are various types of chromatography which differ from one another on the basis of nature of the two phases—mobile and stationary. Various types of chromatography have been listed in Table 1 with the nature of their stationary and mobile phases and basic principle of separation.

Table 1

Name of the chromatography	Stationary phase	Mobile phase	Methods of fixing the stationary phase	Basic principle involved in separation
1. Column	Solid	Liquid	Held in tabular column.	Adsorption
2. Thin Layer (TLC)	Solid	Liquid	Finely divided solid on a glass plate.	Adsorption
3. Gas-Solid (GSC)	Solid	Gas	Held in tabular column.	Adsorption and volatility
4. Partition	Liquid	Liquid	Adsorbed on a porous solid held in a tabular column.	Partition
5. Paper	Liquid	Liquid	Held in pores of a thick paper.	Partition
6. Gas-Liquid (GLC)	Liquid	Gas	Adsorbed on a porous solid held in a tube.	Partition
7. Gel	Liquid	Liquid	Held in the cavities of a polymeric solid.	Molecular sieve

In the present book we shall emphasise on 'Paper Chromatography' avoiding others.

3. PAPER CHROMATOGRAPHY

Paper chromatographic technique was devised by Martin and Synge in the year 1941. It has been widely used for the separation and identification of the components of a mixture especially when it is available in small amounts.

The basic process or principle of separation in paper chromatography is partition of the components of the mixture in two liquids. That is, both the phases, stationary and mobile, in this type of chromatography are liquids. Stationary phase is in the form of liquid (usually water) and is held in the pores of a thick paper (filter paper).

A separation by this technique is achieved by placing a drop of the solution at a point of the strip of paper, and a solvent is allowed to move over the strip by capillary flow. The mobile phase (solvent) is usually saturated with water to prevent dehydration of the paper during the solvent flow. As the mobile phase travels over the initial sample spot, the individual components become distributed between the moving solvent and the water film retained by the material of paper. Under ideal conditions the components migrate at different rates, and separate in different zones just slightly larger than the original spot. The flow of solvent is stopped after travelling some predetermined distance and the position of the separated species is located by visualization process.

4. RETENTION FACTOR OR RETARDATION RATIO

The separated components of the mixture are identified by their R_f values. R_f stands for "Retention factor or Retardation ratio" and can be defined by the expression:

$$R_f = \frac{\text{Distance travelled by component from the origin line}}{\text{Distance travelled by solvent from the origin line}}$$

The determination of R_f values involves measuring the distance from the point of application of the sample to the centre of an individual zone and to the edge of the solvent front. This can be illustrated by the Fig. 1.

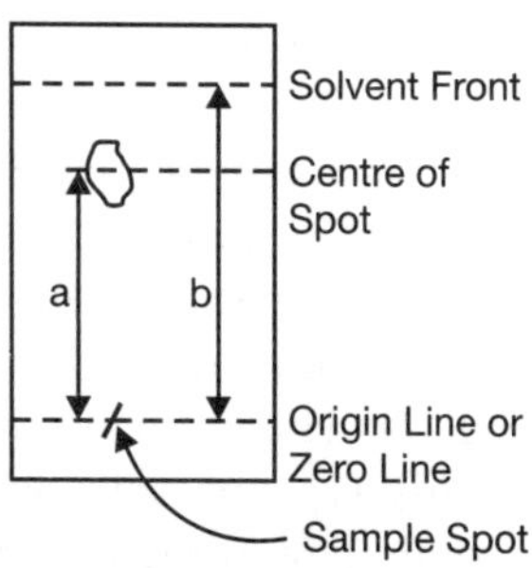

Fig. 1. A developed chromatogram.

In the above diagram, it is seen that a component has travelled a distance, 'a' cm and the solvent has travelled 'b' cm. Thus the R_f value for the component will be

$$R_f = \frac{a}{b}$$

The R_f value of a substance depends upon:

(i) Nature of the mobile phase

(ii) Quality of paper

(iii) Nature of the substance, and

(iv) Temperature.

Things to Remember

(i) The paper strip after running of the solvent and drying is called chromatogram.

(ii) The line at which sample spot is situated, is called origin line or zero line.

(iii) The line upto which the solvent has travelled, is called solvent front.

5. TYPES OF PAPER CHROMATOGRAPHY

There are many types of paper chromatography, viz., Ascending paper chromatography, Descending paper chromatography, Radial paper chromatography, Spiral paper chromatography, Wedge paper chromatography, etc. But depending upon the mode of movement of the mobile phase over the paper, it can be divided into following three types:

5.1. Ascending Paper Chromatography

In this type the mobile phase moves upwards on the paper. That is, the strip of paper is supported at its upper end and the solvent is drawn upon the paper from a reservoir located at the base of chromatographic tank (Fig. 2). The main advantage of this type of development is that it requires no special apparatus. For laboratory work, a wide mouthed test

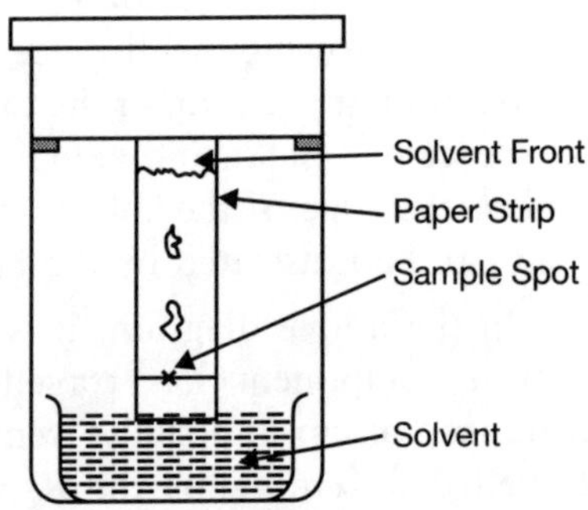

Fig. 2. Ascending technique.

tube or a cylinder may be used which is closed with a cork that has a hook.

5.2. Descending Paper Chromatography

In this the solvent flows from an elevated container down the paper (Fig. 3). That is, the mobile phase descends on the paper.

In paper chromatography, descending technique may be more convenient, though an ascending method can of course also be used.

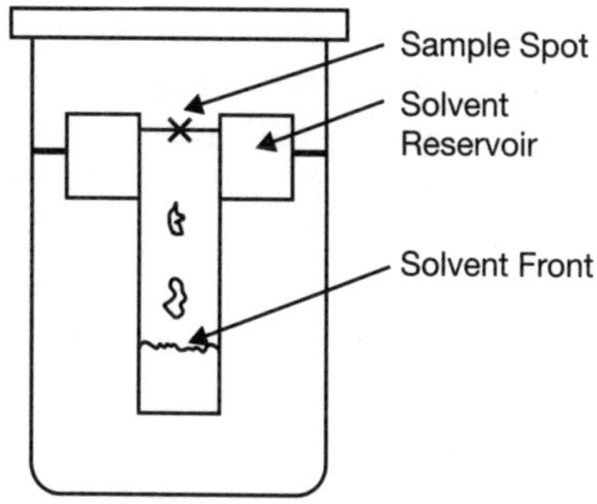

Fig. 3. Descending technique.

For routine work in qualitative analysis, the apparatus depicted in Fig. 4 is inexpensive and efficient. It consists of a glass gas jar or measuring cylinder (say, 50 cm by 7.5 cm), closed with a tight-fitting cork which carries a glass rod. A glass boat is fused

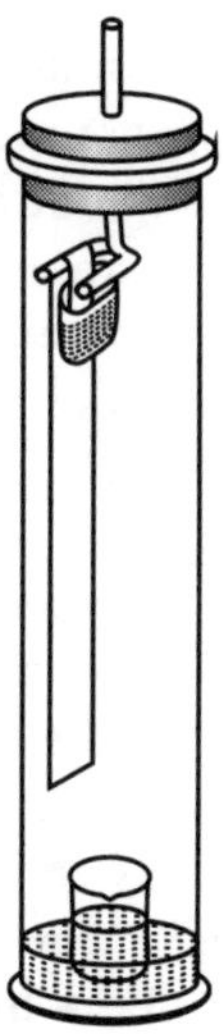

Fig. 4

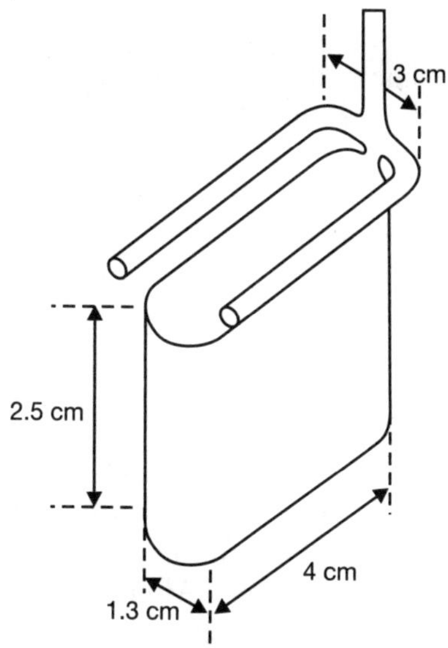

Fig. 5

with this glass rod. The approximate dimensions of the glass boat have been shown in Fig. 5. The boat is suspended about 2.5 cm below the bottom of the cork. It serves as a container for the solvent used in the separations. Two side arms are fused to the glass rod above the glass boat. These act as supports for paper strip. In order to saturate the atmosphere in the cylinder with respect to the solvent, a layer of the latter is kept at the bottom of the glass cylinder. This will prevent evaporation effects from disturbing the equilibrium of the solvent travelling down the paper. In some separations it is also necessary to control the amount of water vapour in the atmosphere. For this purpose a beaker containing water, saturated salt solution or aqueous acid saturated with the solvent is placed on the bottom of the jar. It is advantageous to maintain fairly constant temperature conditions and for some separations the whole apparatus is kept in a thermostatically controlled (±1°) cabinet or bath.

The procedure permits the separation of 1-200 μg of ions. Under favourable conditions as much as 1 mg (1000 μg) of certain ions may be separated by chromatography. A strip of Whatman No. 1 filter paper, 30-45 cm long and 2.5 cm wide, is cut from the rolls supplied by the manufacturers. A light pencil line is drawn across the width of the paper about 8 cm from the upper edge. This acts as zero line. A convenient volume (say, 0.05 ml) of the test solution, containing not more than 1 mg of the mixed metals (say, as chlorides) is applied to the centre of the zero line with the help of a micro pipette (or micro syringe). The strip is held horizontally until the liquid is absorbed. The spot usually forms a wet patch about 2.5 cm^2. This is generally allowed to dry completely in the air (one hour). In some cases the degree of drying is dependent upon the chemical stability of the salts undergoing separation. The strip is then hung vertically in the cylinder with the upper end immersed in the solvent contained in the glass boat. The upper end of the strip may be held down in the glass boat by means of a glass rod. After the solvent has diffused down the paper a sufficient distance to effect a separation (1-24 hours; this may happen when the solvent front has travelled within

2.5-5 cm of the lower edge of the paper), the strip is removed from the cylinder. The position of the solvent front should be marked when the strip is still wet with solvent since it is sometimes difficult to locate it when dry. For some mixed solvents containing water there may be two solvent fronts, one of the dry solvent (lower area—the position of this is noted) and another wet solvent front (upper area) in which the paper is saturated with water. The solvent is then allowed to evaporate, i.e. the strips are dried by pinning or clipping them over a glass rod clamped or supported at a convenient height. The presence and position of the salts on the strip are detected by spraying with a suitable reagent, if the spots of salts are not coloured. The simple type of atomizer is shown in Fig. 6.

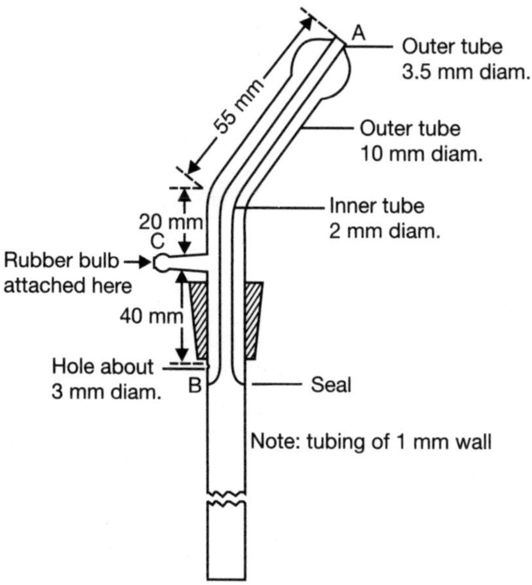

Fig. 6

In the atomizer, a rubber bulb is attached at C. The cork is fitted into a boiling tube or small flask containing suitable reagent for spray. The paper is sprayed first on one side and then on the other.

After spraying, the coloured spots (or bands) may not appear until several hours have passed.

5.3. Horizontal or Radial Paper Chromatography

In this type, the mobile phase moves horizontally over the paper. The paper can be taken in the form of rectangular strip or circular sheet. The name, radial chromatography is justified when the paper is a circular sheet. Radial chromatography is more efficient and recent technique. The sample spot is placed at or near the centre of a circular filter paper. The solvent is fed up to the central spot by a wick dipping in the solvent reservoir (Fig. 7). The solvent spreads out radially, causing the sample components to spread into a series of concentric bands (Fig. 8).

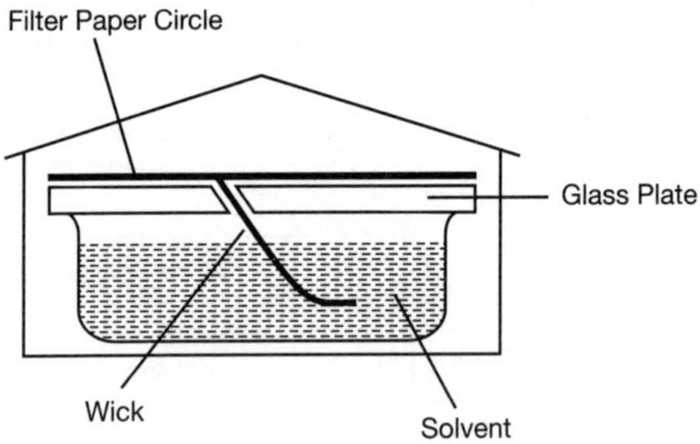

Fig. 7

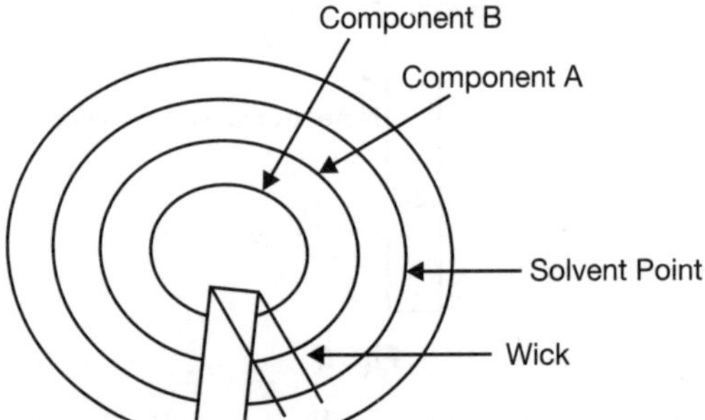

Fig. 8

6. GENERAL EXPERIMENTAL PROCEDURE

Paper chromatography comprises of the following four steps from practical point of view:

(i) Preparation of the filter paper,
(ii) Spotting the paper,
(iii) Development, and
(iv) Identification.

But before proceeding for these steps systematically, first of all, the type of paper chromatography, i.e., ascending, descending or radial, is selected depending on the availability of the apparatus and time and the nature of the materials to be separated.

(i) Preparation of the Filter Paper

A strip of Whatman filter paper No. 1, 3 or 4 is cut about 30-40 cm long (if the height of the jar is about 50 cm) and 1.5 cm wide. Then, a straight-line called origin-line, or start-line or initial line is drawn horizontally along the width of the strip about 3 cm from lower edge (for ascending technique) or 8 cm from the upper edge (for descending technique) with a soft pencil. For radial chromatography, two parallel cuts (2 mm apart) are made from the edge to the centre of 11 cm circle of Whatman filter paper No. 41 as shown in Fig. 8. This piece is bent perpendicular to the circular paper so that it serves as wick.

(ii) Spotting the Paper

A cross mark, "×", is put almost in the middle of the origin-line with a soft pencil. Then the test solution is applied on this cross with a micro pipette or micro syringe. While applying the test solution, care should be taken to see that the spot diameter does not exceed 3 mm. So for applying the test solution, a capillary need only be dipped into the solution and touched to the paper to introduce sufficient solution. The spotted paper is dried in air (1 hour).

(iii) Development

The dry strip is positioned in a chromatographic jar in such a way that its spotted end comes in contact with the mobile phase. The sample spot itself should be about 2 cm apart from the mobile phase.

The mobile phase can be a single substance or a mixture of substances. The composition of the mobile phase (called solvent or developer) is decided by the nature of the species to be separated. The selection of an appropriate mobile phase for a new separation procedure has often been determined by empirical experiments. However, the following points are observed while selecting a suitable solvent (mobile phase):

- The components should be more soluble in the stationary phase as compared to mobile phase.
- The solvent should not react chemically with any of the components or paper.
- It should not interfere with the detection of the spots on the developed chromatogram.
- R_f values of any two components must differ by at least 0.05 in the chosen solvent.
- The composition of solvent system should not change with time.

When the solvent has moved a reasonable distance (say, 15 cm) from the origin line, the strip is removed from the jar and the position of the solvent is marked by drawing a line with a soft pencil.

This line is called solvent front. The paper strip is then dried in air (1-24 hours). It is known as chromatogram.

(iv) Identification

The components of mixture separate in the form of spots lying between the origin-line and the solvent front. Number of components are usually equal to the number of spots appeared. Visualization of the spots is done from their colours. But, if the spots are not coloured then the strip is sprayed (Fig. 6) with a reagent called visualization reagent or spraying reagent. This reagent causes the components to form a derivative which is either coloured or fluoresces when exposed to ultraviolet light.

R_f values also indicate the relative position to be expected for the different components and, therefore, helps in identification of these.

6.1. Solvents Employed

In order to obtain repeatable results, the compositions of the solvents employed must be constant. In some cases, particularly with water-miscible solvents, small variations in the water content may affect considerably the efficiency of the separation. All the solvents used must be of high purity. Some of the solvents employed are detailed below:

(i) n-Butyl alcohol saturated with 3M hydrochloric acid. Equal volumes of the alcohol and 3M hydrochloric acid are shaken together. The upper layer is used.

(ii) Acetylacetone saturated with water. To 7.5 c.c. of the saturated solvent, 0.05 c.c. conc. HCl and 2.5 c.c. dry acetone are added. This solvent is described as acetylacetone saturated with water and containing 0.5 per cent (v/v) of hydrochloric acid (d 1.18) and 25 per cent (v/v) of dry acetone.

(iii) Glacial acetic acid containing 25 per cent (v/v) dry methanol.

(iv) Acetone containing 5 per cent (v/v) water and 8 per cent (v/v) hydrochloric acid (d 1.18).

(v) Pyridine containing 20 per cent (v/v) water and 1 per cent (w/v) potassium thiocyanate.

(vi) Methanol.

(vii) Methyl ethyl ketone containing 30 per cent (v/v) hydrochloric acid (d 1.18).

(viii) Cellosolve (ethylene glycol monoethyl ether) containing 20 per cent (v/v) hydrochloric acid (d 1.18).

(ix) Ethyl ether containing 2 per cent (w/v) dry hydrogen chloride and 7.5 per cent (v/v) dry methanol.

(x) Methyl acetate containing 3 per cent (v/v) methanol and 10 per cent (v/v) water.

(xi) Dry n-butyl alcohol containing 40 per cent (v/v) dry methanol.

(xii) 2-Methyl-tetrahydroxyfuran (tetrahydrosylvan) containing 5 per cent (v/v) water and 10 per cent (v/v) nitric acid (d 1.42).

(xiii) Pyridine containing 10 per cent (v/v) water.

6.2. Purification of Solvents

The following details for the purification of solvents for paper chromatography may be useful:

(i) **Acetone:** It is refluxed with solid potassium hydroxide and potassium permanganate for one hour, and fractionated. The fraction having b.p 56° ± 0.5° is used.

(ii) **Methyl ethyl ketone:** It is refluxed with solid potassium hydroxide and potassium permanganate for one hour, and fractionated. The fraction having b.p. 79.5°-80° is used.

(iii) **Methyl n-propyl ketone:** It is refluxed with solid potassium hydroxide and potassium permanganate for one hour, and fractionated. The fraction having b.p. 99°-102° is used.

(iv) **Methyl alcohol:** The purest commercial (synthetic) alcohol is redistilled and the fraction having b.p. 63° is used.

(v) **n-Butyl alcohol:** It is dried over anhydrous calcium sulphate, filtered and distilled. The fraction having b.p. 116.5°-117.5° is used.

(vi) **Pyridine:** The AnalaR brand is satisfactory.

(vii) **Ethyl acetate:** It is shaken with saturated calcium chloride solution, dried with anhydrous calcium sulphate, filtered and distilled. The fraction having b.p. 76.5°-77.5° is used.

(viii) **Methyl acetate:** It is shaken with saturated calcium chloride solution, dried with anhydrous calcium sulphate, filtered and distilled. The fraction having b.p. 56.5°-57.5° is used.

(ix) **Diethyl ether:** It is refluxed with alkaline potassium permanganate, dried with anhydrous calcium sulphate, filtered and distilled. The fraction having b.p. 35° is used. The ether is free from peroxides.

(x) **2-Methyl-tetrahydroxyfuran (tetrahydrosylvan):** It is refluxed with alkaline potassium permanganate, dried with anhydrous calcium sulphate, filtered and distilled. The fraction having b.p. 79°-80° is used.

6.3. Visualization or Spraying Reagent

A large number of organic and inorganic reagents have been used as visualization or spraying reagent. The criteria applied in selecting a reagent for paper chromatography differ from those usually employed in choosing a 'spot' test reagent. It is not necessary for the reagent to be specific for a certain ion, but it is desirable to have one which will give a test for as many ions as possible. The reagents employed include diphenyl thiocarbazone (dithizone), rubeanic acid, diphenylcarbazide, alizarin, salicylaldoxime, morin, potassium hexacyanoferrate(II), potassium chromate, ammonium sulphide, and hydrogen sulphide (as the free gas). In many instances mixtures of two or more of these reagents are advantageous.

7. PROCEDURES FOR SELECTED PAPER CHROMATOGRAPHIC SEPARATIONS

A number of procedures for separation of coloured components, different cations, and different anions from their mixtures are given below.

7.1. Separation of the Coloured Components Present in the Flower Juice by Ascending Technique

Theory: Flower juice contains organic compounds. Thus, these will be covalent in nature and therefore soluble in organic solvents. Hence the mobile phase should be an organic solvent with some degree of polarity, e.g., a mixture of acetone and HCl.

Apparatus: Gas jar with cover or wide mouthed glass bottle, Whatman filter paper No. 1, capillary tube with fine tip, mortar and pestle.

Chemicals: Acetone, hydrochloric acid, distilled water, flower petals.

Procedure: (i) With the ruler, the height of the gas jar or bottle is measured and strip cut from Whatman filter paper No. 1 almost equal to it (height of the jar) and 5 cm wide. A line about 3 cm from one end of the paper strip is drawn with the help of soft pencil [Fig. 9(b)]. The paper strip is folded in the middle so that the two halves are at an angle to each other [Fig. 9(c)].

(ii) Few petals of coloured flower and about 1 c.c. of acetone are taken in a mortar and pestle and crushed. Now the coloured solution so obtained is separated by decantation.

(iii) With the help of a fine capillary, a small spot of the extract solution is put at about the middle of the pencil line. The spot is dried in air. Sometimes, extract obtained is quite dilute. So it is better to apply several drops on the same spot allowing one to dry before applying the next so that the spot does not spread too far.

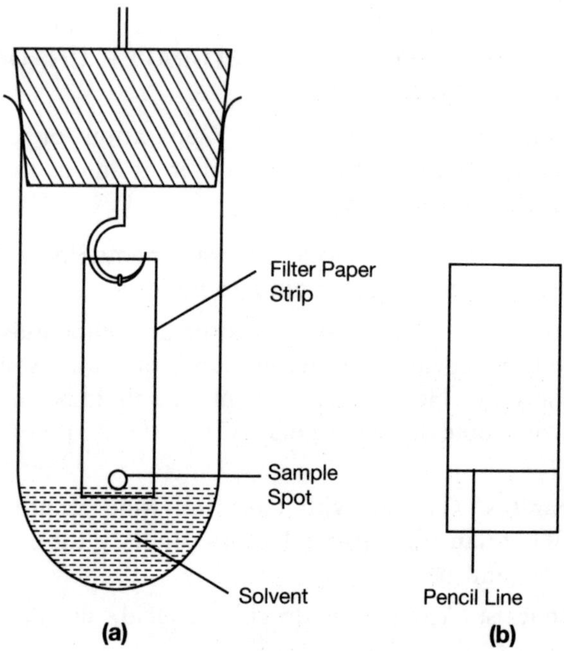

Fig. 9

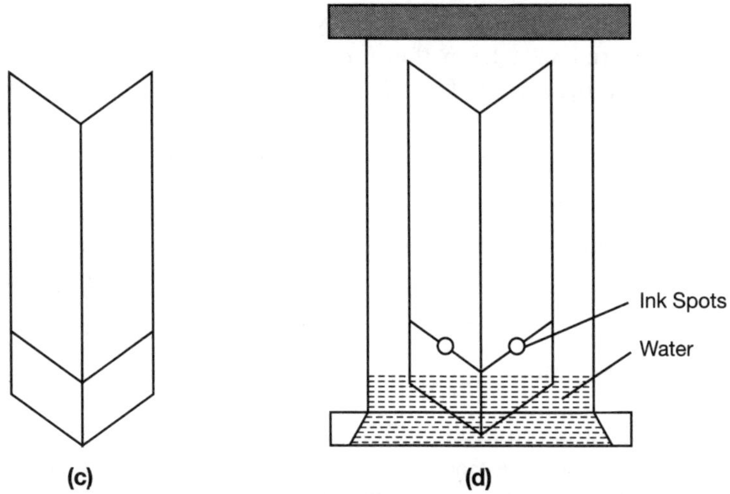

Fig. 9

(iv) Mobile phase in this case is a mixture of 5 c.c. acetone, 1 c.c. dil. HCl, and 5 c.c. water. If acetone is not available, then 5 c.c. dil. HCl and 20 c.c. water can be used as the mobile phase. The mobile phase is poured into the jar or bottle to a depth of about 2 cm and the folded paper strip with the extract spot is put into it in such a way that it stands straight and does not touch the sides of the jar or bottle as shown in Fig. 9(d). If the jar is not available, then the strip is hanged vertically with the help of a hook fixed in a cork fitted into the mouth of a big and wide boiling tube containing the mobile phase as shown in Fig. 9(a).

(v) The jar is covered with a glass cover or thick paper and left undisturbed till the mobile phase reaches almost up to the other end of the strip or rises at least 10-12 cm above the pencil line.

(vi) The strip is taken out of the jar and the position of the solvent front marked with a soft pencil.

(vii) The paper strip is dried and a mark put with pencil around each coloured spot obtained.

(viii) The distances travelled by the solvent and centre of each coloured spot from the line on which spot was applied initially are measured. Then R_f values are calculated for each spot.

(ix) Chromatogram is attached with the result.

Observations: Name of the flower = ________________

Developing solvent or mobile phase = 5 c.c. acetone
+ 1 c.c. dil. HCl
+ 5 c.c. water

Running time = __________

Table 2

Sl. No.	Colour of the spot	Distance of spot from the origin line (a cm)	Distance of solvent front from the origin line (b cm)	$R_f = \frac{a}{b}$
1.	......	......	......	......
2.	......	......	......	......
3.	......	......	......	......
4.	......	......	......	......

Result: No. of coloured components present in the extract of flower = ...

Things to Remember

(i) The experiment procedure has been described here for ascending technique. It can also be done by descending method (see section 5.2).

(ii) A very fine capillary should be used for applying the spot. Diameter of the spot should not exceed 3 mm.

(iii) The distance of the spot from the end should not be more than 3 cm.

(iv) The spot should be allowed to dry before putting another drop on it.

(v) The spot must be dried in air before keeping the strip in the bottle or jar containing the solvent. It may take maximum of 1 hour.

(vi) The strip should not be dried in oven or over flame.

(vii) The spot should not dip in the mobile phase.

(viii) The strip should not touch the wall of the bottle or jar.

(ix) The jar or bottle should not be disturbed after putting the strip in it.

(x) The jar or bottle should almost be air tight.

(xi) In order to get good result, the jar is saturated with vapours of the mobile phase. This can be done by shaking the jar up and down after pouring in the mobile phase.

7.2. Separation of the Mixture of Two Dyes by Ascending Technique

Apparatus: Gas jar with cover or wide mouthed glass bottle, Whatman filter paper No. 1, capillary tube with a fine tip.

Chemicals: Alcohol, distilled water, a mixture of red and blue inks.

Procedure: (i) A strip of Whatman filter paper No. 1 is cut about 1.5 cm wide and length almost equal to that of the jar or bottle.

(ii) A line is drawn about 3 cm from one end of the strip with a pencil.

(iii) A drop of the mixture of red and blue inks is put on the strip almost in the middle of the pencil line with the help of a fine capillary. The spot is dried. The process is repeated 2-3 times putting every drop on the same spot.

(iv) The mobile phase (50% ethanol) is poured in the jar.

(v) The paper strip is suspended vertically in the jar with the help of a hook or glass rod, so that its pencil line end just dips into the mobile phase.

(vi) The jar is covered and kept undisturbed till the mobile phase runs at least 10-15 cm above the pencil line.

(vii) The paper strip is removed and a line drawn at the solvent front. Then it is dried.

(viii) Centres of red and blue spots are marked by the pencil. Then R_f values are determined by measuring distances of the spots and solvent front from zero line.

Observations: Mobile phase = 50% ethanol.

Table 3

Sl. No.	Colour of the spot	Distance of spot from the origin line (cm)	Distance of solvent front from the origin line (cm)	R_f
1.	Red	a_1	b	a_1/b
2.	Blue	a_2	b	a_2/b

Result: R_f value of the red ink = a_1/b

R_f value of the blue ink = a_2/b

Things to Remember

(i) The experimental procedure has been described here for ascending technique. It can also be done by descending technique (see section 5.2).

(ii) Precautions should be taken same as in section 7.1.

7.3. Separation and Identification of Ag^+, Pb^{2+} and Hg_2^{2+} (Group I) by Descending Technique

Apparatus: Gas jar, Whatman filter paper No. 1, capillary tube with a fine tip.

Chemicals: Aqueous solution of mixture of silver nitrate, lead nitrate and mercurous nitrate, dil. HNO_3, mobile phase (n-butyl alcohol, glacial acetic acid, water), visualization reagent (dithizone in chloroform).

Procedure: (i) A strip of Whatman filter paper No. 1 is cut about 1.5 cm wide and length almost equal to the jar or bottle.

(ii) A line is drawn about 8 cm from the upper end of the strip with a pencil.

(iii) The metals are present as nitrates in dil. nitric acid. A drop of the solution is put on the middle of the line drawn. The spot is dried in air and the process is repeated 2-3 times putting every drop on the same spot.

(iv) The mobile phase consists of n-butyl alcohol, mixed with 5 per cent (v/v) glacial acetic acid, followed by water to turbidity. It is taken in the jar and shaken. This is done to make the atmoshere of the jar saturated with the solvent (mobile phase).

(v) Mobile phase is also taken in a glass boat (see section 5.2).

(vi) The upper end of the strip is dipped in the mobile phase contained in a glass boat and the elution (separation) is allowed for 12-16 hours.

(vii) The strip is removed from the extraction vessel. Then a line is drawn for solvent front. Now it is dried in air.

(viii) The strip is sprayed with a 0.05 per cent solution of dithizone in chloroform. In this stage, we get three coloured spots.

(ix) The distances of these coloured spots and solvent front are measured from the zero line. With the help of these distances (and hence R_f values), the given metal ions can be identified by matching the observed R_f values with given R_f values.

Observations:

Mobile phase – n-butyl alcohol + 5 per cent glacial acetic acid + water to turbidity

Visualization reagent – 0.05 per cent dithizone in chloroform.

Table 4

Name of cations	Colour of the spot	Distance of spot from the origin line (cm)	Distance of solvent front from the origin line (cm)	R_f
Pb^{2+}	Rose pink	.....	.....	0.08
Ag^{+}	Orange	.....	.....	0.16
Hg_2^{2+}	Pink	.....	.....	0.85

Result: Cations present in the mixture = ...
....
....

Things to Remember

(i) The experimental procedure has been described here for descending technique. It can also be done by ascending technique (see section 7.1).

(ii) The paper chromatogram may also be developed by exposure to hydrogen sulphide gas, but the results are not as satisfactory as with dithizone.

7.4. Separation and Identification of Hg^{2+}, Pb^{2+}, Bi^{3+}, Cu^{2+} and Cd^{2+} (Group IIA) by Descending Technique

Apparatus: Gas jar, Whatman filter paper No. 1, capillary tube with a fine tip.

Chemicals: Solvent phase (n-butyl alcohol, 3M hydrochloric acid), visualization reagent (dithizone in chloroform and aqueous solution of rhodizonic acid), metal chlorides, dil. HCl.

Procedure: (i) A strip of Whatman filter paper No. 1 is cut about 1.5 cm wide and length almost equal to the jar or bottle.

(ii) A line is drawn about 8 cm from the upper end of the strip with a pencil.

(iii) The metals are present as chlorides in dil. HCl. A drop of the solution is put on the middle of the line drawn. The spot is dried in air for one hour and the process is repeated 2-3 times putting every drop on the same spot.

(iv) The solvent consists of n-butyl alcohol saturated with 3M hydrochloric acid. It is taken in the jar and shaken. This is done to make the atmosphere of the jar saturated with the solvent.

(v) Solvent is also taken in a glass boat (see section 5.2).

(vi) The upper end of the strip is dipped in mobile phase contained in the glass boat and the elution (separation) is allowed for 15-18 hours.

(vii) The strip is removed from the extraction vessel. Then a line is drawn for solvent front. Now it is dried in air.

(viii) The strip is sprayed with a solution of dithizone in chloroform (0.05 per cent w/v).

(ix) Lead gives only a weak colour with dithizone and is best detected by spraying the top portion of the strip with an aqueous solution of rhodizonic acid. Thus five coloured spots appear.

(x) The distances of these coloured spots and solvent front are measured from the zero line. With the help of these distances and hence R_f values, the given metal ions can be identified by matching the observed R_f values with given R_f values.

Observations: Mobile phase — n-butyl alcohol saturated with 3M hydrochloric acid.

Visualization reagent — Dithizone in chloroform and aqueous solution of rhodizonic acid.

Table 5

Name of cations	Name of visualization reagent	Colour of the spot	Distance of spot from the origin line (cm)	Distance of solvent front from the origin line (cm)	R_f
Cu^{2+}	Dithizone	Purple brown	.....	.....	0.20
Pb^{2+}	Rhodizonic acid	Bright blue	.....	.....	0.27
Bi^{3+}	Dithizone	Purple	.....	.....	0.59
Cd^{2+}	Dithizone	Purple	.....	.....	0.77
Hg^{2+}	Dithizone	Pink	.....	.....	0.81

Result: Cations present in the mixture = ...

....

....

....

....

Things to Remember

(i) The experimental procedure given above is descending technique. It can also be done by ascending technique (see section 7.1).

(ii) To obtain a good separation of copper and lead ions, the strip should be at least 45 cm long.

7.5. Separation and Identification of As^{3+}, Sb^{3+}, and Sn^{2+} (Group IIB) by Descending Technique

Apparatus: Gas jar, Whatman filter paper No. 1, capillary tube with a fine tip.

Chemicals: Solvent (acetylacetone, water, conc. HCl, acetone), visualization reagent (dithizone in chloroform), chlorides of arsenic, antimony and tin(II) in dil. HCl.

Procedure: (i) A strip of Whatman filter paper No. 1 is cut about 1.5 cm wide and length almost equal to the jar or bottle.

(ii) A line is drawn about 8 cm from the upper end of the strip with a pencil.

(iii) The metals are present as their chlorides in dilute (2-4M) hydrochloric acid. A drop of the solution is put on the middle of the line drawn. The spot is dried in air for 15 minutes or more and the process is repeated 2-3 times putting every drop on the same spot.

(iv) The solvent consists of 7.5 c.c. acetylacetone (b.p. 137°-141°) saturated with water and treated with 0.05 c.c. conc. HCl and 2.5 c.c. acetone (sufficient of the last named to give a clear solution). It is taken in jar and shaken. This is done to make the atmosphere of the jar saturated with the solvent (mobile phase).

(v) Solvent is also taken in the glass boat (see section 5.2).

(vi) The upper end of the strip is dipped in solvent contained in a glass boat and the elution (separation) is allowed for 1 hour. Movement of solvent is by about 15 cm. The complexes formed are very stable, particularly that with tin.

(vii) The strip is removed from the extraction vessel. Then a line is drawn for solvent front. Now it is dried in air for several minutes.

(viii) The strip is sprayed (before it is completely dry) with spraying reagent—dithizone in chloroform (0.005 per cent w/v). Then it is allowed to dry thoroughly. In this stage, we get three coloured spots.

(ix) The distances of these coloured spots and solvent front are measured from the zero line. With the help of these distances and hence R_f values, the given metal ions can be identified by matching the observed R_f values with given R_f values.

Observations: Mobile phase —7.5 c.c. acetylacetone
+ water
+ 0.05 c.c. conc. HCl
+ 2.5 c.c. acetone

Visualization reagent — Dithizone in chloroform
(0.005 per cent w/v)

Table 6

Name of cations	Colour of the spot	Distance of spot (cm)	Distance of solvent front (cm)	R_f
As^{3+}	Yellow	.....	.....	0.2
Sb^{3+}	Red	.....	.....	0.5
Sn^{2+}	Purple	.....	.....	1.0

Result: Cations present in the mixture = ...
....
....

Things to Remember

(i) The above experiment has been shown by descending technique. It can also be done by ascending technique (see section 7.1).

(ii) Tin is found in the solvent front.

7.6. Separation and Identification of Fe^{3+}, Al^{3+} and Cr^{3+} (Group III) by Descending Technique

Apparatus: Gas jar, Whatman filter paper No. 1, capillary tube with a fine tip.

Chemicals: Solvent (mixture of 3 vol. of glacial acetic acid and 1 vol. of methyl alcohol), visualization reagent (saturated alcoholic solution of alizarin made alkaline by vapour of ammonia and 0.05 per cent solution of benzidine in 2M acetic acid with 0.5M aqueous sodium peroxide), potassium carbonate, chlorides of iron, aluminium and chromium.

Procedure: (i) A strip of Whatman filter paper No. 1 is cut about 1.5 cm wide and length almost equal to the jar or bottle.

(ii) A line is drawn about 8 cm from the upper end of the strip with a pencil.

(iii) The metals are present as chlorides in 5M hydrochloric acid. The test solution is spotted upon filter paper and allowed to evaporate to dryness in air.

(iv) The solvent employed is the mixture of 3 volume of glacial acetic acid and 1 volume (v/v) of dry methanol. It is taken in the jar and shaken. This is done to make the atmosphere of the jar saturated with the solvent-vapour.

(v) A beaker containing saturated solution of potassium carbonate is put in the jar (see Fig. 4) to maintain low humidity. It improves the separation.

(vi) The solvent is taken in a glass boat (see section 5.2).

(vii) The upper end of the strip is dipped in solvent contained in a glass boat and the elution (separation) is allowed to proceed for 12 hours.

(viii) The strip is removed from the extraction vessel. Then a line is drawn for solvent front.

(ix) After the solvent has evaporated, the strip is cut lengthwise in two portions.

(x) One portion is sprayed with a saturated alcoholic solution of alizarin, made alkaline by exposure to ammonia vapour and then warmed. Aluminium appears as red band well separated from a purple band due to iron.

(xi) The other portion of the strip is sprayed first with 0.5M aqueous sodium peroxide and then with a 0.05 per cent solution of benzidine in 2M acetic acid. Chromium is indicated as a bright blue band just behind aluminium. (Danger: The Reagent is carcinogenic.)

Things to Remember

(i) This experiment can also be performed by ascending technique.

(ii) As the reagent is carcinogenic (mentioned above), this experiment should be avoided.

7.7. Separation and Identification of Ni^{2+}, Co^{2+}, Mn^{2+} and Zn^{2+} (Group IV) by Descending Technique

Apparatus: Gas jar, Whatman filter paper No. 1, capillary tube with fine tip.

Chemicals: Solvent (acetone containing 5 per cent water and 8 per cent hydrochloric acid), visualization reagent (ammonia vapour followed by saturated alcoholic solution of alizarin containing 0.1 per cent rubeanic acid and 1 per cent salicylaldoxime), metal chlorides in dil. HCl.

Procedure: (i) A strip of Whatman filter paper No. 1 is cut about 1.5 cm wide and length almost equal to the jar or bottle.

(ii) A line is drawn about 8 cm from the upper end of the strip with a pencil.

(iii) The metals are present as chlorides in dilute hydrochloric acid. A drop of the solution is put on the middle of the line drawn. The spot is dried in air and the process is repeated 2-3 times putting every drop on the same spot.

(iv) The mobile phase consists of acetone containing 5 per cent (v/v) water and 8 per cent (v/v) hydrochloric acid (d 1.18). It is taken in the jar and shaken to saturate the atmosphere of the jar with the vapours of the mobile phase.

(v) The mobile phase is taken in the the glass boat (see section 5.2).

(vi) The upper end of the strip is dipped in mobile phase contained in a glass boat and the elution (separation) is allowed until the mobile phase moves to 25 cm.

(vii) The strip is removed, dried and a line drawn for solvent front.

(viii) Paper strip is exposed to ammonia vapour and then sprayed with a saturated alcoholic solution of alizarin containing 0.1 per cent rubeanic acid and 1 per cent salicylaldoxime. In this stage, we get four coloured spots.

(ix) The distances of these coloured spots and solvent front are measured from the zero line. With the help of these distances and hence R_f values, the given metal ions can be identified by matching the observed R_f values with given R_f values.

Observations: Mobile phase — Acetone + 5 per cent water + 8 per cent hydrochloric acid.

Visualization reagent — Ammonia vapour + saturated alcoholic solution of alizarin containing 0.1 per cent rubeanic acid and 1 per cent salicylaldoxime.

Table 7

Name of cations	Colour of the spot	Distance of spot from the origin line (cm)	Distance of solvent front from the origin line (cm)	R_f
Ni^{2+}	Blue	.....	.....	0.07
Mn^{2+}	Brown	.....	.....	0.3
Co^{2+}	Brown	.....	.....	0.6
Zn^{2+}	Purple	.....	.....	0.9

Result: Cations present in the mixture = ...

....

....

....

Things to Remember

The above experiment can also be done by ascending technique (see section 7.1).

7.8. Separation and Identification of Ca^{2+}, Sr^{2+} and Ba^{2+} (Group V) by Descending Technique

Apparatus: Gas jar, Whatman filter paper No. 1, capillary tube with a fine tip.

Chemicals: Mobile phase (pyridine containing 20 per cent water and 1 per cent potassium thiocyanate), visualization reagent (saturated alcoholic solution of alizarin and 0.5 per cent aqueous sodium rhodizonate), ammonium nitrate or chloride, chlorides of metals.

Procedure: (i) A strip of Whatman filter paper No. 1 is cut about 1.5 cm wide and length almost equal to the jar or bottle.

(ii) A line is drawn about 8 cm from the upper end of the strip with a pencil.

(iii) Metals are present as chlorides. A drop of the solution is put on the middle of the line drawn. The spot is dried in air and the process is repeated 2-3 times putting every drop on the same spot.

(iv) The mobile phase consists of pyridine containing 20 per cent (v/v) water and 1 per cent (w/v) potassium thiocyanate. Pyridine is added to gas jar and shaken to saturate the jar with pyridine vapour. A beaker containing saturated aqueous solution of ammonium nitrate or chloride is put in the jar to make the humidity of the jar between 65 and 80 per cent. Now the mobile phase (solvent) is taken in the glass boat (see Fig. 4).

(v) The upper end of the strip is dipped in mobile phase contained in a glass boat and the elution (separation) is allowed for 5-6 hours. The solvent front moves about 20 cm.

(vi) The strip is removed and dried. A line is drawn for solvent front.

(vii) The strip is sprayed with a saturated alcoholic solution of alizarin. Here orange spot for calcium is seen.

(viii) Further, the strip is sprayed with 0.5 per cent (w/v) aqueous sodium rhodizonate (freshly prepared). It gives reddish violet spot for strontium and pale red spot for barium.

(ix) The distances of these coloured spots and solvent front are measured from the zero line. With the help of these distances and hence R_f values, the given metal ions can be identified by matching the observed R_f values with given R_f values.

Observations: Mobile phase — Pyridine + 20 per cent (v/v) water + 1 per cent (w/v) KSCN.

Visualization reagent — Saturated alcoholic solution of alizarin and 0.5 per cent (w/v) aqueous sodium rhodizonate.

Table 8

Name of cations	Name of visualization reagent	Colour of the spot	Distance of spot (cm)	Distance of solvent front (cm)	R_f
Ca^{2+}	Alizarin	Orange	.....	.....	0.95
Sr^{2+}	Sodium rhodizonate	Reddish violet	.....	.....	0.75
Ba^{2+}	Sodium rhodizonate	Pale red	.....	.....	0.15

Result: Cations present in the mixture = ...
....
....

Things to Remember

The above experiment can also be done by ascending technique (see section 7.1).

7.9. Separation and Identification of Na^+, K^+ and Li^+ (Group VI) by Descending Technique

Apparatus: Gas jar, Whatman filter paper No. 1, capillary tube with a fine tip.

Chemicals: Mobile phase (methanol), visualization reagent (0.1M silver nitrate and saturated fluorescein in 50 per cent alcohol), chlorides of metals.

Procedure: (i) A strip of Whatman filter paper No. 1 is cut about 1.5 cm wide and length almost equal to the jar or bottle.

(ii) A line is drawn about 8 cm from the upper end of the strip with a soft pencil.

(iii) The metals are present as chlorides in neutral solution. The solvent used is methanol. It is taken in the jar and shaken to saturate the jar with solvent vapour.

(iv) The solvent is taken in the glass boat (see Fig. 4).

(v) The upper end of the strip is dipped in the solvent contained in a glass boat and the elution (separation) is allowed to proceed for hours.

(vi) The strip is removed from the extraction vessel and a line is drawn for solvent front. Then it is dried in air.

(vii) The position of the alkali metal chloride bands is detected by spraying with 0.1M silver nitrate and saturated fluorescein in 50 per cent alcohol, and then drying the strip.

(viii) The distances of the bands and solvent front are measured from zero line. With the help of these distances and hence R_f values, the given metal ions can be identified by matching the observed R_f values with given R_f values.

Observations: Mobile phase — Methanol.

Visualization reagent — 0.1M $AgNO_3$ and saturated fluorescein in 50 per cent alcohol.

Table 9

Name of metal ions	Colour of the spot	Distance of band from zero line (cm)	Distance of solvent front from zero line (cm)	R_f
Li^+	.....	.....	.....	0.8
Na^+	.....	.....	.....	0.5
K^+	.....	.....	.....	0.1

Result: Cations present in the mixture = ...

....

....

Things to Remember

The above experiment can also be done by ascending technique (see section 7.1).

7.10. Separation and Identification of Alkaline Earths and Alkali Metals by Descending Technique

The experimental details given here permit the separation and identification of following sets of metal ions:

- Be^{2+}, Mg^{2+}, Ca^{2+}, Sr^{2+}, Ba^{2+}
- Mg^{2+}, Na^{+}, K^{+}
- Mg^{2+}, Ca^{2+}, Na^{+}, K^{+}
- Li^{+}, Na^{+}, K^{+}

Apparatus: Gas jar, Whatman filter paper No. 1, capillary tube with a fine tip.

Chemicals: Solvent (mixture of 80 per cent ethanol and 20 per cent 2M acetic acid), visualization reagent (0.1M aqueous solution of violuric acid), acetates of metals.

Procedure: (i) A strip of Whatman filter paper No. 1 is cut about 1.5 cm wide and length almost equal to the jar or bottle.

(ii) A line is drawn about 8 cm from the upper end of the strip with a soft pencil.

(iii) The metals are present as acetates in dilute acetic acid. A drop of the solution is put on the middle of the line drawn. The spot is dried in air and the process is repeated 2-3 times putting every drop on the same spot.

(iv) The solvent is a mixture of 80 per cent ethanol and 20 per cent (v/v) 2M acetic acid. It is taken in the jar and shaken. This is done to make the atmosphere of the jar saturated with the solvent (mobile phase).

(v) The mobile phase is also taken in a glass boat (see section 5.2).

(vi) The upper end of the strip is dipped in the mobile phase contained in a glass boat and the elution (separation) is allowed for 24 hours.

(vii) The paper strip is removed, the solvent front marked and the strip dried in an air oven at 60°C.

(viii) The strip is sprayed with 0.1M aqueous solution of violuric acid (5-isonitroso-barbituric acid) and then dried at 60°C. This causes the production of coloured spots.

(ix) The distances of these coloured spots and solvent front are measured from the zero line. With the help of these distances and hence R_f values, the given metal ions can be identified by matching the observed R_f values with given R_f values.

Observations: Mobile phase — 80 per cent ethanol + 20 per cent 2M acetic acid

Visualization reagent — 0.1M aqueous solution of violuric acid

Table 10

Name of metal ions	Colour of band	Distance of band from zero line (cm)	Distance of solvent front from zero line (cm)	R_f
Be^{2+}	Yellowish-green	....	....	0.86
Mg^{2+}	Yellowish-red	....	....	0.76
Ca^{2+}	Orange	....	....	0.68
Sr^{2+}	Reddish-violet	....	....	0.55
Ba^{2+}	Pale red	....	....	0.43

Table 11

Name of metal ions	Colour of band	Distance of band from zero line (cm)	Distance of solvent front from zero line (cm)	R_f
Mg^{2+}	Yellowish-red	....	....	0.76
Na^{+}	Reddish-violet	....	....	0.56
K^{+}	Violet	....	....	0.45

Table 12

Name of metal ions	Colour of band	Distance of band from zero line (cm)	Distance of solvent front from zero line (cm)	R_f
Mg^{2+}	Yellowish-red	....	....	0.76
Ca^{2+}	Orange	....	....	0.68
Na^+	Reddish-violet	....	....	0.56
K^+	Violet	....	....	0.45

Table 13

Name of metal ions	Colour of band	Distance of band from zero line (cm)	Distance of solvent front from zero line (cm)	R_f
Li^+	Reddish-violet	....	....	0.76
Na^+	Reddish-violet	....	....	0.56
K^+	Violet	....	....	0.45

Things to Remember

The above experiment can also be performed by ascending method (see section 7.1).

7.11. Separation and Identification of Halides by Descending Technique

Apparatus: Gas jar, Whatman filter paper No. 1, capillary tube with a fine tip.

Chemicals: Solvent (pyridine containing 10 per cent water or acetone containing 20 per cent water), visualization reagent (0.1M $AgNO_3$ + saturated fluorescein in 50 per cent alcohol and zirconium-alizarin reagent), sodium halides.

Procedure: (i) A strip of Whatman filter paper No. 1 is cut about 1.5 cm wide and length almost equal to the jar or bottle.

(ii) A line is drawn about 8 cm from the upper end of the strip with a pencil.

(iii) The halides are present as their sodium salts. A drop of the solution is put on the middle of the line drawn. The spot is dried in air and the process is repeated 2-3 times putting every drop on the same spot.

(iv) The solvent used is either pyridine containing 10 per cent (v/v) water, or acetone containing 20 per cent (v/v) water. It is taken in jar and shaken to saturate the atmosphere of the jar with the solvent-vapour.

(v) The solvent is taken in the glass boat (see Fig. 4) and the upper end of the strip is dipped in the boat.

(vi) The elution is conducted for 1.5-2 hours. Then the strip is removed, dried in air and the solvent front marked with soft pencil.

(vii) After evaporation of the solvent from the strip, the strip is cut longitudinally.

(viii) One part is sprayed with 0.1M $AgNO_3$ and saturated fluorescein in 50 per cent alcohol, and the other is sprayed with the zirconium-alizarin reagent.

(ix) Upon drying, the halides are clearly visible as characteristic bands.

(x) The distances of these coloured spots (bands) and solvent front are measured from the zero line. With the help of these distances and hence R_f values, the given metal ions can be identified by matching the observed R_f values with predetermined R_f values.

Observations: Mobile phase — Pyridine + 10 per cent water or acetone + 20 per cent water.

Visualization reagent — 0.1M $AgNO_3$ + saturated fluorescein in 50 per cent alcohol, and zirconium-alizarin reagent.

Table 14

Name of halide	Name of mobile phase	Distance of band from zero line (cm)	Distance of solvent front from zero line (cm)	R_f
F^-		....	....	0.00
Cl^-	Pyridine	....	....	0.24
Br^-	+10% H_2O	....	....	0.47
I^-		....	....	0.71

Table 15

Name of halide	Name of mobile phase	Distance of band from zero line (cm)	Distance of solvent front from zero line (cm)	R_f
F^-		....	....	0.25
Cl^-	Acetone	....	....	0.51
Br^-	+20% H_2O	....	....	0.62
I^-		....	....	0.77

Things to Remember

The above experiment can also be performed by ascending method (see section 7.1).

8. CHROMATOGRAM FOR DIFFERENT MIXTURES

A chromatogram drawn from an actual photograph is shown below:

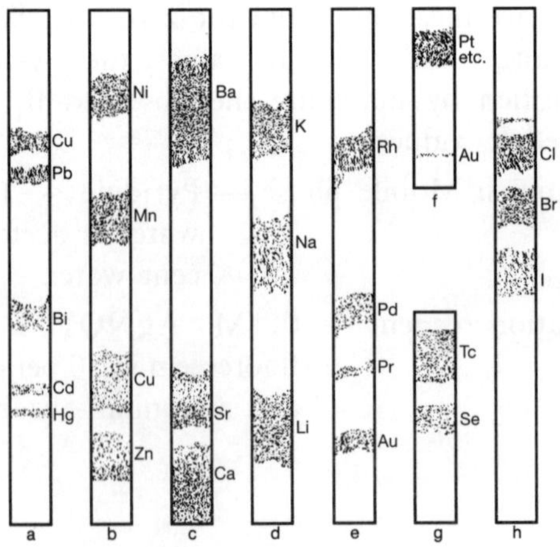

3 INORGANIC PREPARATIONS

1. INTRODUCTION

Preparation of any chemical substance, be it in the laboratory or industry, requires sound knowledge about the underlying chemical changes and skill in handling the apparatus. The experiments regarding preparations of inorganic substances will provide an opportunity to the students to know the functions of various types of apparatus employed by a chemist in the laboratory. The student will also learn the technique of handling the apparatus properly.

While preparing inorganic compounds, a student is expected to follow certain instructions given below:

(i) One should always use the specified quantities of the chemicals required for any preparation.

(ii) One must follow and observe precautions very carefully so as to avoid fires and accidents.

(iii) Whenever any complication (incidental or accidental) arises, please talk to your teacher or demonstrator and do not discuss with your class fellows.

(iv) Whenever the yield is recorded, the product must be dried.

(v) The solution should not be concentrated too much or the solution evaporated to dryness, otherwise the quality of the crystals will be poor.

(vi) The concentrated solution should not be cooled rapidly, otherwise the quality of the crystals will be poor.

(vii) Sometimes it so happens that the concentrated solutions even on cooling do not give crystals (e.g. in case of $CuSO_4$). In such cases a few crystals of the salt should be added in the concentrated solution. This is called "seeding".

(viii) While drying, crystals should be pressed between the folds of a filter paper very gently otherwise these are broken and lose their crystalline shape.

2. PREPARATION OF DOUBLE SALTS

2.1. Preparation of Potash Alum

Theory: Since potash alum is a double salt of potassium sulphate and aluminium sulphate, it is prepared by mixing concentrated solutions containing equimolecular amounts of potassium sulphate, K_2SO_4, and aluminium sulphate, $Al_2(SO_4)_3 \cdot 18H_2O$. The resulting solution is further concentrated to crystallisation point and cooled when octahedral crystals of potash alum appear.

$$K_2SO_4 + Al_2(SO_4)_3 \cdot 18H_2O + 6H_2O \xrightarrow{H^+} \underset{\text{(Potash alum)}}{K_2SO_4 \cdot Al_2(SO_4)_3 \cdot 24H_2O}$$

174 666

Reagents: Potassium sulphate crystals = 2.5 g
Aluminium sulphate crystals = 10.0 g
Conc. H_2SO_4 = 1.0 c.c.

Procedure: (i) 2.5 g potassium sulphate is taken in a clean beaker. 25 c.c. water is added. It is stirred to dissolve the crystals, and heated if necessary. The solution is filtered if it is not clear.

(ii) 10 g aluminium sulphate is taken in another clean beaker. 25 c.c. water is added followed by 1 c.c. conc. H_2SO_4 to prevent hydrolysis of aluminium sulphate. It is heated for

4-5 minutes if milkiness still persists. The solution is filtered if it is not clear.

(iii) These two solutions are mixed in a china dish and the china dish placed on a water bath. It is heated gently. The solution is stirred with a glass rod. The solution is concentrated till a glass rod dipped in the solution deposits a solid crust on it. This indicates that crystallisation point has reached. The dish is covered with a watch glass and the solution allowed to crystallise. Soon octahedral crystals of potash alum separate out.

(iv) Mother liquor is decanted off and the crystals are washed with a small quantity (5 c.c.) of ice cold water.

(v) The crystals are dried by placing them between filter pads or by spreading them over porous plate.

Yield: About 10 g.

2.2. Preparation of Chrome Alum

Theory: The crystals of chrome alum may be prepared by reducing an acidified solution of potassium dichromate. The reduction can be carried out either by using ethyl alcohol or starch solution.

$$K_2Cr_2O_7 \rightarrow \cancel{K_2O} + \cancel{Cr_2O_3} + \cancel{3[O]}$$

$$\cancel{K_2O} + H_2SO_4 \rightarrow K_2SO_4 + H_2O$$

$$\cancel{Cr_2O_3} + 3H_2SO_4 \rightarrow Cr_2(SO_4)_3 + 3H_2O$$

$$\left[CH_3CH_2OH + \cancel{O} \rightarrow CH_3CHO + H_2O\right] \times 3$$

$$\overline{K_2Cr_2O_7 + 4H_2SO_4 + 3C_2H_5OH \rightarrow K_2SO_4 + Cr_2(SO_4)_3 + 7H_2O + 3CH_3CHO}$$

$$K_2SO_4 + Cr_2(SO_4)_3 + 24H_2O \rightarrow K_2SO_4 \cdot Cr_2(SO_4)_3 \cdot 24H_2O$$

(Chrome alum; violet)

Reagents:

Potassium dichromate	= 5.0 g
conc. H_2SO_4	= 4.0 c.c.
Water	= 20.0 c.c.
Alcohol	= 3.0 c.c.
or	
Starch	= 1.5 g

Procedure: (i) 5 g of potassium dichromate is taken in a 200 c.c. beaker. 20 c.c. of water is added to it. Then 4 c.c. of conc. H_2SO_4 is added dropwise and stirred till all the solid dichromate goes into the solution.

(ii) The solution is cooled to 30°C by keeping the beaker in an ice bath. Then 3.0 c.c. of ethyl alcohol is added dropwise with constant stirring (instead of ethyl alcohol, 1.5 g of starch may be added very slowly). One should be careful that the temperature of the solution should not exceed 50°C at any cost; if necessary a small piece of ice may be dropped in the solution to keep the temperature below 50°C.

(iii) The solution is cooled in an ice bath and the contents of the beaker allowed to stand over-night, when crystals of chrome alum separate out. Crystals are separated from the mother liquor, washed with a small amount of cold water, dried between the folds of the filter paper, and weighed.

Yield: 10 g.

2.3. Preparation of Mohr's Salt

Theory: When equimolecular mixture of ferrous sulphate, $FeSO_4 \cdot 7H_2O$ and ammonium sulphate, $(NH_4)_2SO_4$ are dissolved in water and the solution is evaporated, light blue crystals of ferrous ammonium sulphate, $FeSO_4 \cdot (NH_4)_2 SO_4 \cdot 6H_2O$ separate out.

$$FeSO_4 \cdot 7H_2O + (NH_4)_2 SO_4 \rightarrow \underset{\text{(Mohr's salt)}}{FeSO_4 \cdot (NH_4)_2 SO_4 \cdot 6H_2O} + H_2O$$

Reagents: Ferrous sulphate = 28 g
Ammonium sulphate = 14 g
Dil. H_2SO_4 = 2 c.c.

Procedure: (i) 28 g of ferrous sulphate and 14 g of ammonium sulphate are weighed. Both of them are taken in a beaker and about 2 c.c. of dilute H_2SO_4 is added to it to prevent hydrolysis of ferrous sulphate.

(ii) About 100 cc. of distilled water is boiled in another beaker for 2-3 minutes to expel dissolved air. About 50 c.c. of this boiled water is added to the mixture of the salts and stirred with a glass rod, until the salts have completely dissolved. This gives a hot and almost saturated solution of the salts.

(iii) The beaker is covered with a clock glass, and placed in cold water. Cooling water is changed two or three times, and finally replaced with ice water.

(iv) When the contents of the beaker are quite cold and the crystallisation of the salt is complete, the mother liquor is poured off quickly. Then crystals are dried by pressing between the folds of a filter paper or on a porous plate.

Yield: 36 g.

3. PREPARATION OF COMPLEX SALTS

3.1. Preparation of Hexammine Nickel(II) Chloride

Theory: When aqueous solution of nickel chloride is treated with excess of ammonia solution, hexammine nickel(II) chloride is formed.

$$NiCl_2 + 2NH_4OH \rightarrow \underset{\text{(Green ppt.)}}{Ni(OH)_2} + 2NH_4Cl$$

$$Ni(OH)_2 + 2NH_4Cl + 4NH_4OH \rightarrow \underset{\text{[Hexammine nickel(II) chloride]}}{\left[Ni(NH_3)_6\right]Cl_2} + 6H_2O$$

Reagents: Nickel chloride = 12 g

Conc. ammonia = 24 c.c.

Procedure: (i) 12 g of nickel chloride is weighed and taken in a beaker. 20 c.c. of warm water is added and stirred till dissolution. The solution is filtered to remove any insoluble impurities.

(ii) About 24 c.c. of concentrated ammonia is added slowly to the rapidly stirred solution. Green precipitate of nickel hydroxide is formed first. Addition of concentrated ammonia is continued until the green precipitate dissolves.

(iii) The solution is allowed to stand at room temperature for 30 minutes. Crystals of hexammine nickel(II) chloride are obtained. It is filtered on a Buchner funnel. The precipitate is washed with ammonia (density = 0.880; about 50 c.c.); then with acetone and allowed to dry at room temperature in air.

3.2. Preparation of Potassium Chromithiocyanate

Theory: When the saturated solution of potassium thiocyanate is added to the saturated solution of chrome alum, the resulting solution so obtained on heating on a water bath and then on cooling yields dark violet coloured glistering crystals of potassium chromithiocyanate.

$$K_2SO_4 \cdot Cr_2(SO_4)_3 \cdot 24H_2O \rightarrow K_2SO_4 + Cr_2(SO_4)_3 + 24H_2O$$

$$Cr_2(SO_4)_3 + 12KSCN \rightarrow 2K_3[Cr(SCN)_6] + 3K_2SO_4$$

$$K_3[Cr(SCN)_6] + 4H_2O \rightarrow \underset{\text{(Potassium chromithiocyanate)}}{K_3[Cr(SCN)_6] \cdot 4H_2O}$$

Reagents: Potassium thiocyanate = 7.5 g
Chrome alum = 5 g
Ethanol = 20 c.c.

Procedure: (i) 7.5 g of potassium thiocyanate is weighed and taken in a beaker. It is dissolved in minimum volume of water.

(ii) 5 g of chrome alum is weighed and taken in another beaker. It is also dissolved in minimum volume of water.

(iii) These two solutions are mixed together and heated on a water bath at about 80°C for about 60 minutes. The solution is cooled and 20 c.c. of ethanol is added to it to precipitate potassium sulphate. The potassium sulphate is filtered off.

(iv) The filtrate is heated on water bath to make its bulk very small. Now the solution is allowed to stand for cooling whereby dark violet and glistering crystals of potassium chromithiocyanate are obtained. Finally these crystals are recrystallised from ethanol.

Yield: 9 g.

3.3 Preparation of Potassium Trioxalatoaluminate

Theory: When freshly precipitated aluminium hydroxide is boiled with solution of potassium oxalate and oxalic acid followed by crystallisation, crystals of potassium trioxalatoaluminate are formed.

$$2Al(OH)_3 + 3K_2C_2O_4 + 3H_2C_2O_4 \rightarrow \underset{\text{(Pot. trioxalatoaluminate)}}{2K_3\left[Al(C_2O_4)_3\right]} + 6H_2O$$

Reagents: Aluminium sulphate octadecahydrate = 7 g
Potassium oxalate monohydrate = 6 g
Oxalic acid dihydrate = 4 g
Sodium hydroxide = 2.5 g

Procedure: (i) 7 g of aluminium sulphate octadecahydrate is dissolved in 100 c.c. of water. Now aqueous solution of sodium hydroxide (2.5 g in 20 c.c. of water) is added to it with stirring. Consequently, precipitate of aluminium hydroxide is produced. It is filtered and the filtrate is discarded.

(ii) 4 g of oxalic acid dihydrate and 6 g of potassium oxalate monohydrate are dissolved in 100 c.c. of water.

(iii) Freshly precipitated aluminium hydroxide is added to solution of potassium oxalate monohydrate and oxalic acid dihydrate. It is boiled and filtered if any unreacted aluminium hydroxide appears. The filtrate is evaporated for crystallisation.

3.4. Preparation of Potassium Trioxalatochromate(III)

Theory: Potassium trioxalatochromate(III) is prepared by treating potassium dichromate with oxalic acid and potassium oxalate.

In first step, potassium dichromate oxidises oxalic acid and is itself reduced to Cr^{3+}.

$$Cr_2O_7^{2-} + 14H^+ + \cancel{6e} \rightarrow 2Cr^{3+} + 7H_2O$$

$$\left[H_2C_2O_4 \rightarrow 2CO_2 + 2H^+ + \cancel{2e}\right] \times 3$$

$$\overline{Cr_2O_7^{2-} + 3H_2C_2O_4 + 8H^+ \rightarrow 2Cr^{3+} + 7H_2O + 6CO_2 \uparrow}$$

In second step, Cr^{3+} ion reacts with potassium oxalate and oxalic acid to produce potassium trioxalatochromate(III).

$$3K_2C_2O_4 + 3H_2C_2O_4 + 2Cr^{3+} \rightarrow \underset{\text{[Pot. trioxalatochromate(III)]}}{2K_3\left[Cr(C_2O_4)_3\right]} + 6H^+$$

Reagents: Pot. oxalate monohydrate = 3.0 g
Oxalic acid dihydrate = 7.0 g
Potassium dichromate = 2.5 g
Acetone = 10 c.c.

Procedure: (i) 3 g of potassium oxalate and 7 g of oxalic acid are dissolved in 100 c.c. of water contained in a conical flask.

(ii) Now 2.5 g potassium dichromate is added slowly (a pinch at a time) with vigorous shaking.

(iii) After addition of whole amount (2.5 g) of potassium dichromate, the solution is evaporated to crystallisation point. Upon cooling, crystals of potassium trioxalatochromate(III) separate out. It is filtered under suction. Crystals of the complex are washed with acetone and dried.

Yield: 3 g.

3.5. Preparation of Potassium Trioxalatoferrate(III)

Theory: Reaction of ferrous sulphate with oxalic acid, potassium oxalate and hydrogen peroxide causes the formation of potassium trioxalatoferrate(III).

$$FeSO_4 + H_2C_2O_4 \rightarrow FeC_2O_4 + H_2SO_4$$

$$FeC_2O_4 + 2K_2C_2O_4 \rightarrow K_4\left[Fe^{II}(C_2O_4)_3\right]$$

$$2K_4\left[Fe^{II}(C_2O_4)_3\right] + H_2C_2O_4 + H_2O_2$$
$$\rightarrow 2K_3\left[Fe^{III}(C_2O_4)_3\right] + K_2C_2O_4 + 2H_2O$$

$$K_3\left[Fe^{III}(C_2O_4)_3\right] + 3H_2O \rightarrow \underset{\text{[Pot. trioxalatoferrate(III) trihydrate]}}{K_3\left[Fe^{III}(C_2O_4)_3\right]\cdot 3H_2O}$$

Reagents: Mohr's salt,
10% oxalic acid solution,
Potassium oxalate monohydrate,
4N H_2SO_4 solution,
Ethyl alcohol,
Acetone, and
20 volume H_2O_2

Procedure: (i) 15 g of ferrous ammonium sulphate is dissolved in 50 c.c. of hot distilled water. 1 c.c. 4N H_2SO_4 is added to it.

(ii) The solution is stirred using a mechanical or a magnetic stirrer and in the same time 7 c.c. of 10% oxalic acid solution is added to it very slowly.

(iii) Now the solution is heated to boiling with continuous stirring with a glass rod. Consequently, the precipitate of ferrous oxalate is formed. It is allowed to settle down and the supernatant liquid decanted without disturbing the precipitate as far as possible.

(iv) 50 c.c. of hot water is added to the precipitate and stirred well. The precipitate is allowed to settle down and the supernatant liquid is decanted carefully.

(v) 10 g potassium oxalate (monohydrate) is dissolved in 30 c.c. of hot water and slowly added to the precipitate of ferrous oxalate.

(vi) The solution is kept on a water bath. 25 c.c. of 20 volume H_2O_2 is added slowly and continuously with stirring. The temperature of the solution is maintained above 40°C by keeping the solution on a water bath.

(vii) The solution may have traces of precipitate of ferric hydroxide at this stage. Hence the solution is heated to boiling. Now 10 c.c. of 10% oxalic acid solution is added very rapidly to this continuously stirred and boiling solution. Then 3 c.c. more of 10% oxalic acid solution is added drop by drop.

(viii) The hot solution is filtered through a previously weighed sintered bed funnel or crucible No. G3. At this stage, if any crystals are formed, they are dissolved in 30 c.c. of warm ethyl alcohol.

(ix) The filtrate is collected and kept aside for crystallisation in dark.

(x) Bright green coloured crystals are obtained. It is filtered through the weighed sintered bed crucible. Crystals are washed with 50% ethyl alcohol and then with a small quantity of acetone.

(xi) The suction is continued until the crystals become dry. Further, they are dried and weighed.

Things to Remember

The crystals, being extremely photosensitive, should be preserved in dark and exposure to light must be avoided.

3.6. Preparation of Prussian Blue

Theory: When iron is treated with dilute H_2SO_4, ferrous sulphate is formed.

$$Fe + H_2SO_4 \rightarrow FeSO_4 + H_2 \uparrow$$

On addition of potassium ferrocyanide to the solution of ferrous sulphate, turnbull's blue is first formed.

$$2FeSO_4 + K_4\left[Fe^{II}(CN)_6\right] \rightarrow \underset{\text{(Turnbull's blue)}}{Fe_2^{II}\left[Fe^{II}(CN)_6\right]} + 2K_2SO_4$$

When this solution is heated, we get green solution of ferrous sulphate which gets oxidised into ferric sulphate in presence of air.

$$2FeSO_4 + H_2SO_4 + O \rightarrow Fe_2(SO_4)_3 + H_2O$$

This ferric sulphate reacts with potassium ferrocyanide to produce ferric ferrocyanide which is known as prussian blue.

$$2Fe_2^{III}(SO_4)_3 + 3K_4\left[Fe^{II}(CN)_6\right] \rightarrow \underset{\substack{\text{(Ferric ferrocyanide;}\\ \text{prussian blue)}}}{Fe_4^{III}\left[Fe^{II}(CN)_6\right]_3} + 6K_2SO_4$$

Reagents: Iron filings = 0.5 g
Potassium ferrocyanide = 4.5 g
Ethyl alcohol = 15 c.c.
Dil. H_2SO_4 = 50 c.c.

Procedure: (i) Iron filings (0.5 g) are taken in a little water. To this 10 c.c. of dilute H_2SO_4 is added. It is heated on wire gauze. As soon as the reaction becomes vigorous, the burner is removed and the solution is cooled.

(ii) The solution is filtered. If residue remains, 10 c.c. dilute H_2SO_4 is added to it and the process repeated. Filtrate is collected.

(iii) A saturated solution of 4.5 g of potassium ferrocyanide is added to the filtrate. The resulting solution is heated until it acquires green colour. It is now kept in air for nearly two hours. The solution changes to blue.

Things to Remember

The addition of 1-2 c.c. of $K_2Cr_2O_7$ will increase the oxidation of green product.

3.7. Preparation of Reinecke's Salt

Theory: When powdered ammonium thiocyanate is melted and mixed with ammonium dichromate, a violent reaction takes place with the evolution of ammonia. It results in the formation of red plates of Reinecke's salt.

$$(NH_4)_2 Cr_2O_7 \rightarrow Cr_2O_3 + 4H_2O + N_2 \uparrow$$

$$8NH_4SCN + Cr_2O_3 \rightarrow 2\underbrace{NH_4\left[Cr^{III}(NH_3)_2(SCN)_4\right]}_{\substack{\text{Reinecke's salt} \\ \text{[Amm. tetrathiocyanato-} \\ \text{diammine chromate(III)]}}} + 2NH_3 + 3H_2O$$

Reagents: Ammonium thiocyanate = 7.5 g
Ammonium dichromate = 2.5 g
Ethyl alcohol = 20 c.c.

Procedure: (i) Ammonium thiocyanate (7.5 g) is taken in a big silica dish and heated until it begins to melt.

(ii) To this 2.5 g of powdered ammonium dichromate is added. The whole quantity of ammonium dichromate should not be added in bulk but 0.25 g at a time. As soon as ammonium dichromate is added to ammonium thiocyanate, violent reaction takes place with the evolution of ammonia. Silica dish is cooled and the solid mass present in it is powdered.

(iii) This powdered solid mass is the impure sample of Reinecke's salt associated with impurities like ammonium thiocyanate and dichromate. The impure mass is washed with small quantity of ethanol several times. Ethanol removes ammonium thiocyanate by dissolving it. Then the solid mass is washed with cold water. The cold water removes dichromate by dissolving it.

(iv) The remaining purified mass is dissolved in hot water

and kept for crystallisation. Consequently, glistering red plates of Reinecke's salt separate out.

Yield: 7 g.

3.8. Preparation of Sodium Trioxalatoferrate(III)

Theory: Sodium trioxalatoferrate(III) is prepared by dissolving hydrated ferric oxide in oxalic acid and sodium oxalate solution.

$$Fe_2O_3 + 3Na_2C_2O_4 + 3H_2C_2O_4 \rightarrow 2Na_3\left[Fe\left(C_2O_4\right)_3\right] + 3H_2O$$

$$Na_3\left[Fe\left(C_2O_4\right)_3\right] + 9H_2O \rightarrow \underset{\text{[Sodium trioxalato ferrate(III); green crystals]}}{Na_3\left[Fe\left(C_2O_4\right)_3\right] \cdot 9H_2O}$$

Reagents: Ferric chloride = 10 g
Sodium hydroxide = 11.5 g
Oxalic acid = 12 g

Procedure: (i) 10 g of ferric chloride is dissolved in 10 c.c. of water.

(ii) 7.5 g of sodium hydroxide is dissolved in 5 c.c. of water.

(iii) The whole of the sodium hydroxide solution is added to the ferric chloride solution slowly. Consequently, precipitate of ferric hydroxide is obtained. It is filtered and the precipitate is washed with hot water several times.

(iv) In another beaker, 12 g of oxalic acid is dissolved in 50 c.c. of water. To this solution, the pellets of sodium hydroxide (4 g) are added. It causes the formation of sodium oxalate. The resulting solution is heated.

(v) Now hydrated ferric oxide is added to the hot solution of sodium oxalate with constant stirring by a glass rod.

(vi) The solution is filtered and the residue, if any, rejected. The green filtrate is concentrated on a water bath to get the green crystals of sodium ferrioxalate. The crystals are washed first with ice-cold water and then with alcohol. The crystals are dried in a desiccator.

Yield: 15 g.

3.9. Preparation of Tetrammine Copper(II) Sulphate

Theory: It is prepared by adding ammonia to a solution of copper sulphate when cupric hydroxide is precipitated which dissolves in excess of ammonia to give a deep blue solution. The hydrated tetrammine cupric sulphate is then precipitated from the solution by adding ethyl alcohol since it is insoluble in alcohol.

$$CuSO_4 + 4NH_3 + H_2O \rightarrow \left[Cu\left(NH_3\right)_4\right]SO_4 \cdot H_2O$$

[Tetrammine copper(II) sulphate; blue]

Reagents: Copper sulphate = 5 g

Conc. ammonia = 10 c.c.

Ethyl alcohol = 40 c.c.

Procedure: (i) 5 g of crystalline cupric sulphate, $CuSO_4{\cdot}5H_2O$ is dissolved in minimum amount of water. Few drops of conc. H_2SO_4 are also added, if necessary, to clear up the solution.

(ii) Concentrated ammonia solution is added slowly until the pale blue precipitate first formed completely dissolves to give a clear, deep blue solution, and there persists a distinct smell of ammonia.

(iii) Now 40 c.c. ethyl alcohol is added slowly, with constant stirring until the blue colour is nearly discharged. The beaker is heated to 60-70°C with a clock glass over the water bath for about 15-16 minutes. Then the beaker is kept aside and allowed to stand for 1 hour. Consequently, long needle shaped blue crystals of tetrammine cupric sulphate monohydrate separate out. It is filtered and the crystals are washed with a little ethyl alcohol. Crystals are dried in a desiccator.

Yield: 4 g.

3.10. Preparation of Sodium Cobaltinitrite

Theory: When the solution of cobalt nitrate is mixed with sodium nitrite in cold and then this mixture acidified with glacial acetic acid, sodium cobaltinitrite is produced.

$$CH_3COOH + NaNO_2 \rightarrow CH_3COONa + HNO_2$$

$$HNO_2 \rightarrow H^+ + NO_2^-$$

$$Co^{2+} + NO_2^- + 2H^+ \rightarrow Co^{3+} + NO + H_2O$$

$$Co^{3+} + 6NO_2^- \rightarrow \left[Co(NO_2)_6\right]^{3-}$$

$$Co^{2+} + 7NO_2^- + 2H^+ \rightarrow \underset{\text{(Cobaltinitrite; yellow)}}{\left[Co(NO_2)_6\right]^{3-}} + H_2O + NO$$

Reagents: Sodium nitrite = 15 g
Cobalt nitrate = 5 g
Glacial acetic acid = 2.5 c.c.
Ethyl alcohol = 22 c.c.

Procedure: (i) 15 g of sodium nitrite is dissolved in 15 c.c. of warm water.

(ii) The solution is cooled to 50°C. To this solution 5 g cobalt nitrate is added, stirring the solution continuously until the salt is completely dissolved.

(iii) Now acetic acid (2.5 c.c. glacial acetic acid + 2.5 c.c. water) is gradually added to this solution with constant stirring. It is filtered to remove any impurities.

(iv) 20 c.c. of ethyl alcohol is added slowly to the clear solution and the latter kept overnight for the precipitate to settle. The yellow precipitate of sodium cobaltinitrite is obtained on a Buchner funnel and washed with ethyl alcohol until the washings become colourless. The precipitate is allowed to dry in air by spreading it on a filter paper.

Yield: 7 g.

4 VOLUMETRIC ANALYSIS

1. INTRODUCTION

Let us first define quantitative analysis.

"Quantitative analysis is an analysis of the sample which describes the determination of accurate quantity of constituent(s)."

Quantitative analysis is of two types:

(i) Volumetric analysis

(ii) Gravimetric analysis

The chapter is concerned only with volumetric analysis.

"Volumetric analysis is a type of quantitative analysis which describes the determination of accurate quantity of constituent(s) through measurement of volumes of solutions of the interacting substances."

Volumetric analysis has certain advantages over the gravimetric analysis. These are:

(i) Volumetric analysis does not involve laborious operations like precipitation, filtration, drying and ignition.

(ii) Volumetric analysis is rapid and more accurate.

2. TITRATION

The process of volumetric analysis is termed as titration. In titration, two solutions are allowed to react with each other in a conical flask known as titration flask. The strength of one of the solutions is known and the strength of the second solution is to be determined. The solution of known strength is known as standard solution. Suppose, standard solution is taken in a burette and definite volume of solution of unknown strength is taken in a conical flask. Then solution of burette is added to the conical flask dropwise. The process is allowed to continue till completion of the reaction. Now the volume of the standard solution required to react completely with the known volume of the solution taken in the flask is well-known to us. This volume is actually difference between final reading and initial reading of the burette. Thus following three are known to us:

(i) Volume of solution of burette.

(ii) Strength of solution of burette.

(iii) Volume of solution of conical flask.

With the help of above three, we can calculate the strength of solution of conical flask.

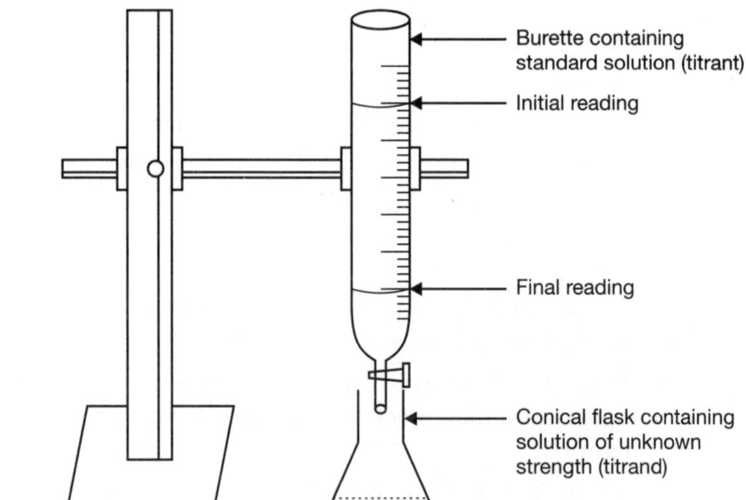

Fig. 1

"Thus the process of finding out the volume of the titrant (standard solution) required to react completely with a known volume of solution under analysis is known as titration."

In titration, we always use a term 'end point' which may be defined as below:

> "The exact point at which a reaction is completed during titration is called its end point or equivalence point."

In order to recognise the end point, certain substances are used. These substances are known as indicators. Thus indicators may be defined as below:

> "Indicators are those substances which indicate the end point by changing their colour on the completion of the reactions."

In the ideal titration, the visible end point will coincide with the theoretical end point. In practice, however, a small difference occurs. This difference is termed as titration error. The choice of indicator should be such that the titration error is too small.

Note: Standard solution may also be taken in the conical flask and solution of unknown strength in the burette.

3. CLASSIFICATION OF VOLUMETRIC ANALYSIS

Volumetric analysis has been classified in five classes:

3.1. Acidimetry and Alkalimetry

In acidimetry and alkalimetry, the strength of an acid is determined by titrating it against a standard alkali and *vice versa*, e.g.

$$HCl + NaOH \rightarrow NaCl + H_2O$$

or, $H^+ + OH^- \rightarrow H_2O$

Determination of strength of an alkali, by titration with standard acid, is known as acidimetry.

Determination of strength of an acid, by titration with standard alkali, is known as alkalimetry.

3.2. Redox Titration

Red̲uction + O̲x̲idation = R̲e̲d̲o̲x̲

In redox titration, reducing agent is estimated by titrating it with a standard solution of oxidising agent and *vice versa*.

Generally, the oxidising agent is employed as a standard solution. For example, oxalic acid can be estimated by titrating it with standard solution of acidified potassium permanganate.

$$\left[MnO_4^- + 8H^+ + 5e \rightarrow Mn^{2+} + 4H_2O\right] \times 2$$

$$\left[C_2O_4^{2-} \rightarrow 2CO_2 + 2e\right] \times 5$$

$$\underset{\text{(Permanganate ion; O.A.)}}{2MnO_4^-} + \underset{\text{(Oxalate ion; R.A.)}}{5C_2O_4^{2-}} + 16H^+ \rightarrow 2Mn^{2+} + 8H_2O + 10CO_2$$

3.3. Iodometry and Iodimetry

These titrations also involve oxidation and reduction reactions but for the sake of simplicity these are studied separately.

Iodimetry is a titration which involves the use of standard solution of iodine (O.A.).

Iodometry is a titration which involves the use of liberated iodine from the reaction between iodide and oxidising agent.

The usefulness of iodine in volumetric analysis depends largely upon its reaction with sodium thiosulphate (R.A.).

$$2S_2O_3^{2-} \rightarrow S_4O_6^{2-} + 2e$$

$$I_2 + 2e \rightarrow 2I^-$$

$$\underset{\text{(Thiosulphate ion)}}{2S_2O_3^{2-}} + I_2 \rightarrow \underset{\text{(Tetrathionate ion)}}{S_4O_6^{2-}} + 2I^-$$

This is really a useful reaction which provides a means of estimating oxidising and reducing agents, e.g.

$$\text{Iodimetry}\begin{cases} As_2O_3 + 2I_2 + 2H_2O \rightarrow As_2O_5 + 4HI \\ 2Na_2S_2O_3 + I_2 \rightarrow Na_2S_4O_6 + 2NaI \end{cases}$$

$$\text{Iodometry}\begin{cases} 2CuSO_4 + 4KI \rightarrow 2K_2SO_4 + Cu_2I_2\downarrow + I_2 \\ 2Na_2S_2O_3 + I_2 \rightarrow Na_2S_4O_6 + 2NaI \end{cases}$$

3.4. Precipitation Titration

In this type of titration, the strength of a solution is determined by its complete precipitation with the help of standard solution of another substance, e.g.

$$AgNO_3 + NaCl \rightarrow \underset{\text{(White ppt.)}}{AgCl \downarrow} + NaNO_3$$

$$AgNO_3 + KSCN \rightarrow \underset{\text{(White ppt.)}}{AgSCN \downarrow} + KNO_3$$

$$2AgNO_3 + K_2CrO_4 \rightarrow \underset{\text{(Brick red ppt.)}}{Ag_2CrO_4 \downarrow} + 2KNO_3$$

The reactions, which involve silver nitrate solution, are called argentimetric reactions.

3.5. Complexometric Titration

In this type of titration, titrand (solution of conical flask) and titrant (solution of burette) react together to form soluble complex ion or undissociated neutral molecule. In such titrations, mainly ethylene diamine tetracetic acid (EDTA) is taken as a complexing agent.

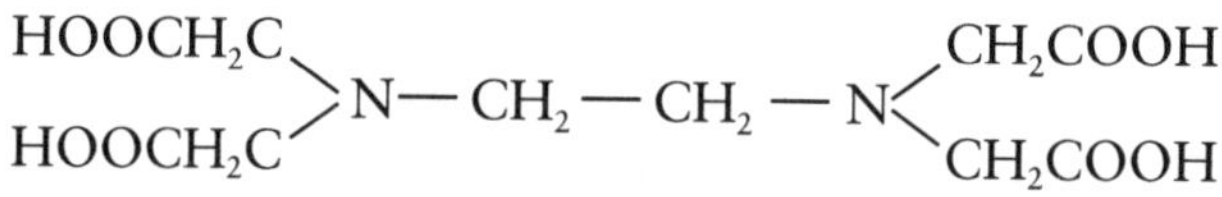

EDTA is hexadentate ligand due to having four carboxyl groups and two nitrogen atoms. Thus it has high chelating power.

The dissociation constants of the hydrogen atoms of the four carboxyl groups suggest that two of these hydrogen atoms are strongly acidic and this leads to Zwitter ion structure of EDTA.

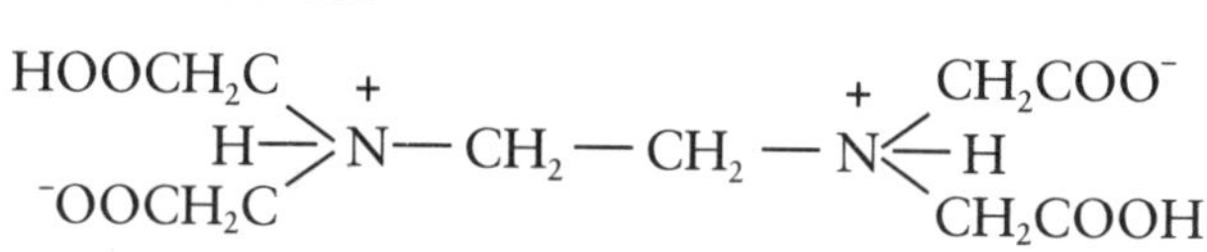

For simplicity, EDTA is written as H_4Y and its disodium salt is written as Na_2H_2Y. In EDTA titration, disodium salt of EDTA is used. It dissociates in water giving anion, H_2Y^{2-}.

$$Na_2H_2Y \rightarrow 2Na^+ + H_2Y^{2-}$$

The anion, H_2Y^{2-} is an excellent complexing agent. It forms 1:1 complexes with nearly all polyvalent metal ions. Its reactions with cations may be written as below:

$$M^{2+} + H_2Y^{2-} \rightarrow MY^{2-} + 2H^+$$

$$M^{3+} + H_2Y^{2-} \rightarrow MY^{-} + 2H^+$$

$$M^{4+} + H_2Y^{2-} \rightarrow MY + 2H^+$$

$$M^{n+} + H_2Y^{2-} \rightarrow MY^{n-4} + 2H^+$$

All metal-EDTA complexes have similar structure, but they differ in charge they carry. One such structure for the complex with divalent metal ion is shown in the figure:

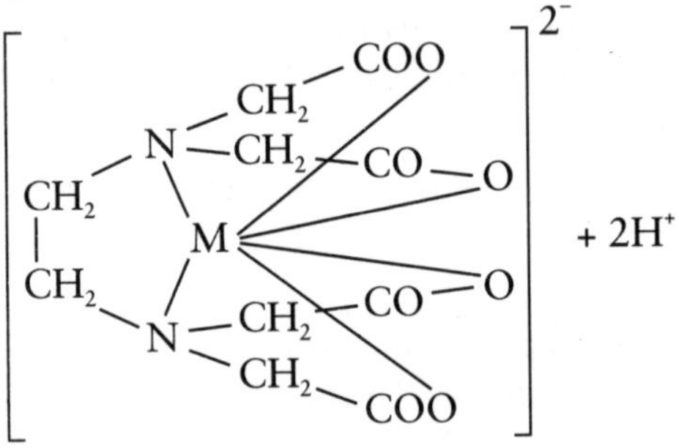

4. STRENGTH OF SOLUTION

The strength of a solution can be expressed in the following terms:

4.1. Normality

The number of gm-equivalents of a solute present in 1 litre of solution is known as normality of that solution. It is expressed as 'N'.

For example, xN NaOH solution means x gm-equivalent of NaOH is dissolved in 1 litre of the solution. Thus normality of this solution will be x.

4.2. Gram Per Litre

The number of grams of solute present in 1 litre of solution is known as gm per litre.

$$\text{Normality} = \frac{\text{Number of grams of solute per litre}}{\text{Equivalent weight}}$$

or, Number of gm of solute per litre = Normality × Eq. wt.

or, Gram per litre = Normality × Eq. wt.

For example, x gm per litre NaOH solution means x gm NaOH is present in 1 litre of the solution.

5. STANDARD SOLUTION

The solution of known strength is known as standard solution. Few standard solutions are discussed below:

5.1. Normal Solution

When 1 litre of solution contains 1 gm-eq. of solute, it is called normal solution or N solution. Normality of this solution will be 1, i.e.

$N = 1$

5.2. Semi Normal Solution

When 1 litre of solution contains $\frac{1}{2}$ gm-eq. of solute, it is called semi normal solution or $\frac{N}{2}$ solution. Normality of this solution will be $\frac{1}{2}$, i.e.

$$N = \frac{1}{2}$$

5.3. Deci Normal Solution

When 1 litre of solution contains $\frac{1}{10}$ gm-eq. of solute, it is called deci normal solution or $\frac{N}{10}$ solution. The normality of this solution will be $\frac{1}{10}$, i.e.,

$$N = \frac{1}{10}$$

5.4. Centi Normal Solution

When 1 litre of solution contains $\frac{1}{100}$ gm-eq. of solute it is called centi normal solution or $\frac{N}{100}$ solution. The normality of this solution will be $\frac{1}{100}$, i.e.

$$N = \frac{1}{100}$$

6. STANDARDISATION

Determination of exact strength of a roughly standard solution by titrating it with a standard solution of primary standard substance is known as standardisation.

For example, we can determine the exact strength of roughly deci normal solution of $KMnO_4$ by titrating it with $\frac{N}{10}$ oxalic acid (primary standard substance). We can do so by assigning factor (f) to roughly $\frac{N}{10}$ $KMnO_4$ solution.

Suppose, 25 c.c. $\frac{N}{10}$ oxalic acid requires 23.7 c.c. $\frac{N}{10}$ (f = x) $KMnO_4$ solution for its complete oxidation, i.e.,

$$23.7 \text{ c.c. } \frac{N}{10}(f = x)\ KMnO_4 = 25 \text{ c.c. } \frac{N}{10} \text{ oxalic acid}$$

$$V_1N_1 = V_2N_2$$

$$23.7 \times \frac{N}{10} \times x = 25 \times \frac{N}{10}$$

$$\text{or,}\quad x = \frac{25}{23.7}$$

$$\text{or,}\quad x = 1.05$$

Therefore, exact strength of $KMnO_4$ solution

$$= \frac{N}{10}(f = 1.05)$$

7. EQUIVALENT WEIGHT

Equivalent weight of a substance is that weight of the substance which is chemically equivalent to 1 gm-atom of hydrogen, $\frac{1}{2}$ gm-atom of oxygen or 1 gm-atom of chlorine.

When equivalent weight of a substance is expressed in gram, it is called gm-equivalent weight or gm-equivalent.

7.1. Equivalent Weight of an Acid

$$\text{Eq. wt. of an acid} = \frac{\text{Molecular wt. of an acid}}{\text{Number of replaceable hydrogen ions present in one molecule}}$$

$$= \frac{\text{Molecular wt. of an acid}}{\text{Basicity of an acid}}$$

7.2. Equivalent Weight of a Base

$$\text{Eq. wt. of a base} = \frac{\text{Molecular weight of a base}}{\text{Number of replaceable hydroxide ions present in one molecule}}$$

$$= \frac{\text{Molecular weight of a base}}{\text{Acidity of a base}}$$

7.3. Equivalent Weight of a Salt

$$\text{Eq. wt. of a salt} = \frac{\text{Molecular wt. of a salt}}{\text{Total number of positive or negative valencies of its radicals}}$$

7.4. Equivalent Weight of an Oxidising Agent

Equivalent wt. of an oxidising agent is equal to its molecular weight divided by no. of electrons absorbed by its one molecule.

$$\text{Eq. wt. of an O.A.} = \frac{\text{Molecular weight of an O.A.}}{\text{Number of electrons absorbed by one molecule of the O.A.}}$$

Equivalent wt. of $KMnO_4$ in acidic medium:

Acidified potassium permanganate ($KMnO_4/H_2SO_4$) oxidises a reducing agent and gets reduced itself into manganous ion (Mn^{2+}).

$$MnO_4^- + 8H^+ + 5e \rightarrow Mn^{2+} + 4H_2O$$

$$\text{Eq. wt. of } KMnO_4 = \frac{\text{Molecular weight of } KMnO_4}{\text{Number of electrons absorbed by one molecule of } KMnO_4}$$

$$= \frac{39 + 55 + 64}{5}$$

$$= \frac{158}{5}$$

$$= 31.6$$

Equivalent wt. of $KMnO_4$ in neutral medium:

Aqueous solution of potassium permanganate oxidises a reducing agent and gets reduced itself into manganese dioxide.

$$MnO_4^- + 2H_2O + 3e \rightarrow MnO_2 + 4OH^-$$

$$\text{Eq. wt. of } KMnO_4 = \frac{\text{Molecular wt. of } KMnO_4}{\text{Number of electrons absorbed by one molecule of } KMnO_4}$$

$$= \frac{39 + 55 + 64}{3}$$

$$= \frac{158}{3}$$

$$= 52.6$$

Equivalent weight of $KMnO_4$ in alkaline medium:

Alkaline potassium permanganate ($KMnO_4/OH^-$) oxidises a reducing agent and gets reduced itself into manganate ion.

$$MnO_4^- + e \rightarrow MnO_4^{2-}$$

$$\text{Eq. wt. of } KMnO_4 = \frac{\text{Molecular weight of } KMnO_4}{\text{Number of electrons absorbed by one molecule of } KMnO_4}$$

$$= \frac{39 + 55 + 64}{1}$$

$$= \frac{158}{1}$$

$$= 158$$

Equivalent wt. of $K_2Cr_2O_7$:

Acidified potassium dichromate oxidises a reducing agent and gets reduced into Cr^{3+} ion.

$$Cr_2O_7^{2-} + 14H^+ + 6e \rightarrow 2Cr^{3+} + 7H_2O$$

$$\text{Eq. wt. of } K_2Cr_2O_7 = \frac{\text{Molecular weight of } K_2Cr_2O_7}{\text{Number of electrons absorbed by one molecule of } K_2Cr_2O_7}$$

$$= \frac{39 \times 2 + 52 \times 2 + 16 \times 7}{6}$$

$$= \frac{78 + 104 + 112}{6}$$

$$= \frac{294}{6}$$

$$= 49$$

Equivalent weight of $CuSO_4{\cdot}5H_2O$:

Copper sulphate oxidises potassium iodide into iodine and gets reduced into cuprous iodide.

$$Cu^{2+} + e \rightarrow Cu^+$$

Eq. wt. of $CuSO_4{\cdot}5H_2O$

$$= \frac{\text{Molecular weight of } CuSO_4{\cdot}5H_2O}{\text{Number of electrons absorbed by one molecule of } CuSO_4{\cdot}5H_2O}$$

$$= \frac{63.5 + 32 + 64 + 90}{1}$$

$$= \frac{249.5}{1}$$

$$= 249.5$$

7.5. Equivalent Weight of a Reducing Agent

Equivalent weight of a reducing agent is equal to its molecular weight divided by number of electrons released by one molecule of it.

$$\text{Eq. wt. of a R.A.} = \frac{\text{Molecular wt. of a R.A.}}{\text{Number of electrons released by one molecule of a R.A.}}$$

Equivalent wt. of Mohr's salt:

Ferrous ion of Mohr's salt reduces an oxidising agent and gets oxidised itself into ferric ion.

$$Fe^{2+} \rightarrow Fe^{3+} + e$$

Eq. wt. of $FeSO_4{\cdot}(NH_4)_2\,SO_4{\cdot}6H_2O$

$$= \frac{\text{Molecular wt. of Mohr's salt}}{\text{Number of electrons released by one molecule of Mohr's salt}}$$

$$= \frac{392}{1}$$

$$= 392$$

Equivalent wt. of sodium thiosulphate:

Sodium thiosulphate reduces iodine into iodide and gets oxidised itself into sodium tetrathionate.

$$2S_2O_3^{2-} \rightarrow S_4O_6^{2-} + 2e$$

$$2S_2O_3^{2-} \equiv 2e$$

$$\therefore\ S_2O_3^{2-} \equiv e$$

Eq. wt. of $Na_2S_2O_3{\cdot}5H_2O$

$$= \frac{\text{Molecular wt. of } Na_2S_2O_3{\cdot}5H_2O}{\text{Number of electrons released by one molecule of sod. thiosulphate}}$$

$$= \frac{46 + 64 + 48 + 90}{1}$$

$$= 248$$

Equivalent wt. of arsenious oxide:

Arsenious oxide reduces iodine into iodide and gets oxidised itself into arsenic oxide.

$$As_2O_3 + 2H_2O \rightarrow As_2O_5 + 4H^+ + 4e$$

$$\text{Eq. wt. of } As_2O_3 = \frac{\text{Molecular wt. of } As_2O_3}{\text{Number of electrons released by one molecule of } As_2O_3}$$

$$= \frac{75 \times 2 + 16 \times 3}{4}$$

$$= \frac{150 + 48}{4}$$

$$= \frac{198}{4}$$

$$= 49.5$$

Equivalent wt. of oxalic acid:

Oxalic acid reduces an oxidising agent and gets oxidised itself into carbon dioxide.

$$C_2O_4^{2-} \rightarrow 2CO_2 + 2e$$

Eq. wt. of $H_2C_2O_4 \cdot 2H_2O$

$$= \frac{\text{Molecular wt. of } H_2C_2O_4 \cdot 2H_2O}{\text{Number of electrons released by one molecule of oxalic acid}}$$

$$= \frac{126}{2}$$

$$= 63$$

8. PRIMARY STANDARD SUBSTANCE

A primary standard substance possesses the following characteristics:

(i) It must be obtained in a high degree of purity.

(ii) It should be stable and unaffected by the atmosphere.

(iii) It should not be deliquescent or efflorescent.

(iv) It should be readily soluble in water.

(v) Its solution in distilled water should remain well on keeping.

Crystalline oxalic acid, anhydrous sodium carbonate, KCl, $AgNO_3$, Mohr's salt, etc. are some of the common primary standard substances which are used in the laboratory.

A substance which does not possess the above characteristics is called a secondary standard substance, e.g., HCl, H_2SO_4, HNO_3, NaOH, KOH, $KMnO_4$, etc.

9. CALCULATION OF STRENGTH BASED ON NORMALITY

The strength (gm per litre) of a solution can be determined by knowing its normality through titration. We know that 1 gm-eq. of an acid neutralises 1 gm-eq. of a base, e.g.,

1000 c.c. N acid ≡ 1000 c.c. N base

When two solutions (one acid and another base) of different normalities react with one another, their interacting volumes will be inversely proportional to their normalities, e.g.,

(10 c.c.) (2N) acid ≡ (2 c.c.) (10N) base

or, 10 × 2 c.c. N acid ≡ 2 × 10 c.c. N base

or, 20 c.c. N acid ≡ 20 c.c. N base

Let V_1 c.c. of an acid having normality N_1 reacts completely with V_2 c.c. of a base having normality N_2.

For acid:

$$V_1 \propto \frac{1}{N_1}$$

or, $$V_1 = \frac{K}{N_1} \qquad \text{... (1)}$$

where K is constant.

For base:

$$V_2 \propto \frac{1}{N_2}$$

or, $V_2 = \dfrac{K}{N_2}$ ... (2)

where K is constant.

Dividing (1) by (2)

$$\frac{V_1}{V_2} = \frac{N_2}{N_1}$$

or, $V_1N_1 = V_2N_2$ ... (3)

or, Volume of acid × Normality of acid

= Volume of base × Normality of base

The equation (3) is known as normality equation. By using this equation, we can calculate the normality of a solution if volumes of two solutions and normality of second solution are known. If we know the equivalent weight of the substance, then the strength (gm per litre) of the solution may be calculated by the following relation:

Gram per litre = Normality × Eq. wt.

10. INDICATOR

An indicator can be defined as under:

- An indicator is a substance which indicates the end point generally by a change of colour of the solution.
- An indicator is a substance which indicates physico-chemical condition of a chemical reaction.
- An indicator is a substance which is used for the visual detection and determination of a specific constituent present in any substance or mixture of substances.

Indicator plays an important role in the volumetric analysis. The success of a volumetric exercise depends upon the indicator. When the reaction between two solutions is completed, then the slight excess of one solution is revealed by the colour change in the solution. This stage of reaction is known as end point.

We have three kinds of indicators with respect to the use of indicators.

10.1. Internal Indicator

These are chemical substances which are added to the volumetric flask in which the titration is carried out, e.g., phenolphthalein, methyl orange, starch, etc.

10.2. External Indicator

These indicators cannot be added to the volumetric flask in which the titration is carried out, e.g., potassium ferricyanide in the titration of potassium dichromate with ferrous ammonium sulphate. External indicators are really very few.

Potassium ferricyanide gives blue colour with $FeSO_4$ solution and light brown colour with ferric ions.

$$2K_3\left[Fe(CN)_6\right] + 3FeSO_4 \rightarrow \underset{\text{(Blue colour)}}{Fe_3\left[Fe(CN)_6\right]_2} + 3K_2SO_4$$

The potassium dichromate solution is run in from the burette into a solution of ferrous sulphate and drops of the reaction mixture are occasionally removed and touched with the drop of the indicator placed on a porcelain tile. The formation of blue colour indicates that end point has not been reached. In the beginning, when only a small volume of potassium dichromate has been added from the burette, the colour produced with indicator drops is intense blue and which goes on fading with the addition of dichromate solution, till when the reaction is over, the blue colour is not visible. With series of titrations the correct end point in located.

Another example is provided by the titration of zinc ions with standard potassium ferrocyanide solution. Here a solution of uranyl acetate or nitrate is the external indicator and titration is continued until a drop of the solution just imparts a brown colour to the indicator. External indicators are being superseded by the more satisfactory internal indicators.

10.3. Self Indicator

If the reagent is strongly coloured and is decolourised by the substance that is being determined or it is converted into slightly coloured compounds, the end point is marked by the colour of a slight excess of reagent. This is well illustrated by potassium permanganate. One drop (0.05 c.c.) of excess of say 0.1 N potassium permanganate will impact a visible pink colouration to several hundred c.c. of solution, even in the presence of slightly coloured ions, such as ferric ions. The colour of ceric sulphate and of iodine solutions have also been employed in the detection of end points. But the colour change is not so marked as for potassium permanganate.

11. INTERNAL INDICATOR

Internal indicators may be divided according to the types of reactions in which they are used.

11.1. Acid-base Indicator

An indicator for the detection of the end point of acid-base titration is a substance that undergoes visible changes within a limited range of hydrogen ion concentration.

The indicators used in acidimetry and alkalimetry are either weak organic acids (indicator acids) or weak organic bases (indicator bases), the dissociated form of which has a colour different from that of the undissociated form. A basic indicator possesses a coloured cation while an acid indicator possesses a coloured anion.

Phenolphthalein (HPh):

(i) Ostwald's Theory: It is a weak acid and is mostly unionised. The unionised molecules are colourless while on ionisation it gives colourless H^+ ions and pink coloured Ph^- ions.

$$\underset{\text{(Undissociated form; colourless)}}{HPh} \rightleftharpoons \underset{\text{(Colourless)}}{H^+} + \underset{\text{(Pink)}}{Ph^-} \quad \begin{pmatrix}\text{Dissociation} \\ \text{very small}\end{pmatrix}$$

Addition of H^+ ions i.e. addition of acid suppresses the ionisation of HPh due to the presence of common H^+ ions and the solution remains colourless. When a base like NaOH or KOH is added, the OH^- ions produced from them combine with H^+ ions obtained from phenolphthalein to form unionised water. The removal of H^+ ions causes the dissociation of HPh in the right direction i.e. more of HPh will dissociate to produce Ph^- ions.

$$HPh \rightleftharpoons H^+ + Ph^-$$

$$NaOH \rightleftharpoons OH^- + Na^+$$

$$\begin{matrix} OH^- & & Na^+ \\ \Updownarrow & & \Updownarrow \\ \underset{\text{(Feebly ionised)}}{H_2O} & & \underset{\text{(Highly ionised)}}{NaPh} \end{matrix}$$

The colour produced in the alkaline medium is due to the presence of Ph^- ions.

$$\underset{\text{(Acid medium)}}{HPh} \rightleftharpoons \underset{\text{(Alkaline medium)}}{H^+ + Ph^-}$$

(ii) **Quinonoid Theory:** The colour change of phenolphthalein can also be explained on the basis of Quinonoid Theory. Phenolphthalein can exist in two tautomeric forms having different colours, one can exist in acid medium and another in alkaline medium. The two forms possess different structures and are known as benzenoid and quinonoid.

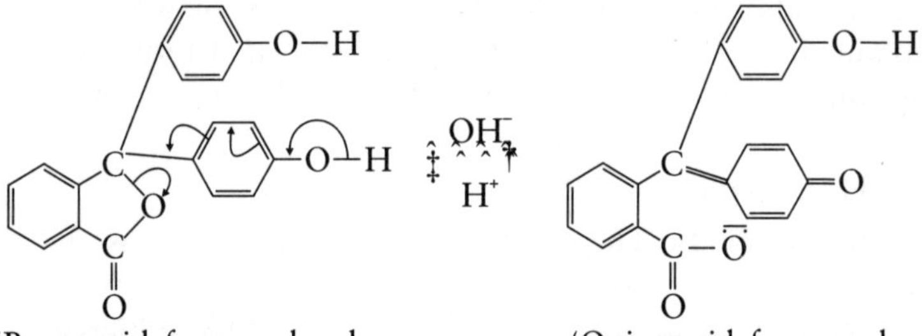

(Benzenoid form; colourless in acid medium) (Quinonoid form; red in alkaline medium)

Methyl Orange (MeOH):

(i) **Ostwald's Theory:** It is a weak base and ionises as:

$$\underset{\text{(Undissociated form; yellow)}}{MeOH} \rightleftharpoons \underset{\text{(Red)}}{Me^+} + \underset{\text{(Colourless)}}{OH^-} \quad \left(\begin{matrix}\text{Dissociation}\\ \text{very small}\end{matrix}\right)$$

Methyl orange is not dissociated in the presence of the bases like NaOH and KOH due to the presence of common OH^- ions. Therefore, the solution remains yellow. But when an acid like HCl is added to the solution it produces H^+ ions which combine with OH^- ions to form feebly ionised water molecules. This increases the ionisation of methyl orange thus producing more Me^+ ions in solution. Hence the solution changes from yellow to pink in the presence of acids.

$$MeOH \rightleftharpoons Me^+ + OH^-$$

$$HCl \rightleftharpoons Cl^- + H^+$$

$$Cl^- \rightleftharpoons \underset{\text{(Highly ionised)}}{MeCl} \qquad H^+ \rightleftharpoons \underset{\text{(Feebly ionised)}}{H_2O}$$

The colour produced in the acid medium is due to presence of Me^+ ions.

$$\underset{\text{(Alkaline medium)}}{MeOH} \rightleftharpoons \underset{\text{(Acid medium)}}{Me^+} + OH^-$$

(ii) Quinonoid Theory: The colour change of methyl orange can also be explained on the basis of Quinonoid Theory. Methyl orange can exist in two tautomeric forms having different colours, one can exist in acid medium and another in alkaline medium. The two forms possess different structures and are known as benzenoid and quinonoid. The colour change is due to the interconversion between these two forms.

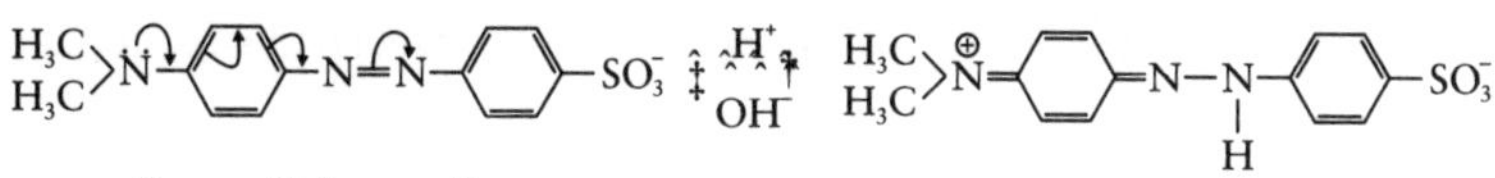

(Benzenoid form; yellow in alkaline medium) (Quinonoid form; red in acid medium)

11.1.1. Transition interval of indicator

Transition interval of indicator is also known as pH range of indicator. Every acid-base indicator has a definite range of pH over which it changes its colour. The usual range of colour change is approximately two pH units.

Table 1: Most Common Acid-base Indicators

Indicator	pH range	Colour in acid	Colour in alkali
Methyl orange	3 – 4.5	Red	Yellow
Methyl red	4.5 – 6.5	Red	Yellow
Phenolphthalein	8 – 10	Colourless	Red

11.1.2. Choice of indicator

A suitable indicator for an acid-base titration is that which has its pH range close to the pH change of the solution at equivalent point.

Table 2: Choice of Indicator

Titration	pH change at equilibrium	Suitable indicator
Strong acid and strong base	3 – 10	Methyl orange, methyl red and phenolphthalein
Strong acid and weak base	3.5 – 7	Methyl orange and methyl red
Weak acid and strong base	6.5 – 10	Phenolphthalein
Weak acid and weak base	6 – 8 (This change is not large. Hence indicator changes its colour slowly and thus no sharp point is visible)	No indicator is suitable

11.2. Redox Indicator

The acid-base indicator is employed to mark the sudden change in pH during acid-base titration. Similarly a redox indicator marks the sudden change in electrode potential (reduction potential) of the titre (solution to be titrated) at equivalence point.

Redox indicator is an organic dye. Its colour depends upon electrode potential of the medium. An equilibrium exists between oxidised and reduced forms of indicator.

$$In_{ox} + ne \rightleftharpoons In_{red}$$

where, In_{ox} = Oxidised form of indicator.

In_{red} = Reduced form of indicator.

Thus a potential is produced due to this equilibrium. According to Nernst's equation, the electrode potential of this system is given by

$$E = E^\circ + \frac{0.0591}{n} \cdot \log_{10} \frac{[In_{ox}]}{[In_{red}]} \quad \text{(at 25°C)} \qquad \ldots (1)$$

The oxidised form and the reduced form of an indicator have different colours.

Case I: When $\frac{[In_{ox}]}{[In_{red}]} > 10$,

then the indicator shows the colour of oxidised form.

Case II: When $\frac{[In_{red}]}{[In_{ox}]} > 10$,

then indicator shows the colour of reduced form.

Thus the colour of indicator depends upon the value of $\frac{[In_{ox}]}{[In_{red}]}$. From equation (1), it is obvious that the value of $\frac{[In_{ox}]}{[In_{red}]}$ depends upon the electrode potential (E). Thus the colour of redox indicator depends upon the electrode potential of the system in which it is placed.

11.2.1. Choice of redox indicator

The choice of suitable indicator for a redox titration is determined by the electrode potential value in the equivalence point region and that of the indicator itself. Ideally E°_{In} should coincide the electrode potential at the equivalence point.

The value of electrode potential at the equivalence point (E_{ep}) is calculated with the help of following relation:

$$E_{ep} = \frac{n_1 E_1^\circ + n_2 E_2^\circ}{n_1 + n_2}$$

where n_1 and n_2 are the number of electrons involved in the redox reactions of the titre and titrant system respectively.

Example:

E° of $Fe^{3+}/Fe^{2+} = 0.77$ volt $= E_1^\circ$

E° of $Cr_2O_7^{2-}/Cr^{3+} = 1.33$ volt $= E_2^\circ$

$Fe^{3+} + e \rightarrow Fe^{2+}$

$n = 1$

$Cr_2O_7^{2-} + 14H^+ + 6e \rightarrow 2Cr^{3+} + 7H_2O$

$n_2 = 6$

$$\therefore \quad E_{ep} = \frac{n_1E_1^\circ + n_2E_2^\circ}{n_1 + n_2}$$

$$= \frac{1 \times 0.77 + 6 \times 1.33}{1 + 6}$$

$$= \frac{0.77 + 7.98}{7}$$

$$= \frac{8.75}{7}$$

$$= 1.25 \text{ volt}$$

Barium diphenyl amine sulphonate ($E^\circ = 0.83$ volt) is generally used as indicator in the titration of Fe^{2+} ion with dichromate solution.

Here $E^\circ_{In} = 0.83$ volt while $E_{ep} = 1.25$ volt. Since E°_{In} does not coincide with E°_{ep} this indicator does not give good result. In order to get good result with this indicator, the electrode potential of Fe^{3+}/Fe^{2+} system is decreased by converting Fe^{3+} ions into complex ions by ortho phosphoric acid.

$$Fe^{3+} + H_3PO_4 \rightleftharpoons [Fe(HPO_4)]^+ + 2H^+$$

Exact value of E°_{ep} is not known for this system due to indefinite structure of the complex ion.

Table 3: List of Redox Indicators

Indicator	Colour change		E°
	Ox	Red	at pH = 3
Indigomonosulphonate	Blue	Colourless	0.25
Diphenylamine (Diphenylbenzidine)	Violet	Colourless	0.76±0.1
Methylene blue	Blue	Colourless	0.52
Diphenylamine sulphonic acid	Red violet	Colourless	0.85

11.2.2. Diphenylamine and diphenylbenzidine

Diphenylamine and diphenylbenzidine are very slightly soluble in water. Stock solutions are, therefore, prepared in sulphuric acid. Knop was the first to propose the use of diphenylamine as an indicator in the titration of ferrous ion with potassium dichromate.

Upon oxidation, diphenylamine and diphenylbenzidine give a beautiful violet oxidation product which is called diphenylbenzidine violet.

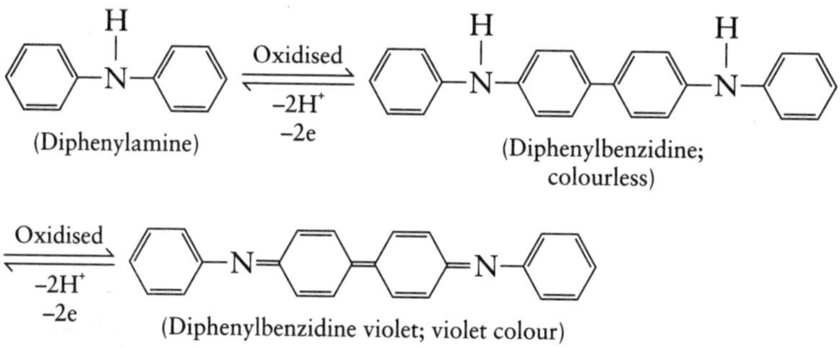

The action of diphenylamine as an indicator depends upon its oxidation first into colourless diphenylbenzidine which is the real indicator and is reversibly further oxidised to diphenylbenzidine violet.

Things to Remember

The electrode potential written above always reveals reduction potential.

11.3. Precipitation Indicator

Precipitation indicator is a substance which produces coloured precipitate at the end point.

Potassium Chromate:

In Mohr's method (1856), solution of halide is taken in a conical flask and few drops of 2% solution of potassium chromate are added to it. Standard $AgNO_3$ solution is taken in a burette and added dropwise into the conical flask. Silver nitrate reacts with halide to form a precipitate of silver halide.

$$Ag^+ + Cl^- \rightarrow \underset{\text{(White ppt.)}}{AgCl \downarrow}$$

In beginning silver nitrate does not react with potassium chromate to form brick red precipitate of silver chromate. This is so because AgCl is more insoluble than Ag_2CrO_4. In other words we can say that solubility product of AgCl is less than that of Ag_2CrO_4.

$$S_{AgCl} = 1.2 \times 10^{-12}$$

$$S_{Ag_2CrO_4} = 9 \times 10^{-12}$$

As soon as all chloride ions of the conical flask are consumed, further addition of $AgNO_3$ solution causes the precipitation of brick red coloured silver chromate. This stage is end point.

$$2Ag^+ + CrO_4^{2-} \rightarrow \underset{\text{(Brick red ppt.)}}{Ag_2CrO_4 \downarrow}$$

Ferric Alum:

In Volhard's method (1874), silver nitrate solution is taken in a conical flask and few drops of aqueous solution of ferric alum, $K_2SO_4 \cdot Fe_2(SO_4)_3 \cdot 24H_2O$ are added. Potassium thiocyanate or ammonium thiocyanate solution is taken in the burette and added dropwise into the conical flask. Potassium thiocyanate reacts with silver ion to form white precipitate of silver thiocyanate.

$$Ag^+ + SCN^- \rightarrow \underset{\text{(White ppt.)}}{AgSCN \downarrow}$$

In beginning potassium thiocyanate does not react with ferric ion of the indicator because silver thiocyanate is more insoluble than ferric thiocyanate complex. As soon as all the silver ions of the conical flask are consumed, further addition of potassium thiocyanate causes the precipitation of red coloured ferric thiocyanate complex. This stage is end point.

$$Fe^{3+} + 3SCN^- \rightarrow \underset{\text{(Red ppt.)}}{Fe(SCN)_3} \downarrow$$

Things to Remember

(i) In Volhard's method, the conical flask must contain free nitric acid which prevents hydrolysis of ferric ion of indicator.

(ii) In Volhard's method, the conical flask should be shaken vigorously. It is so because silver thiocyanate (ppt.) has a strong tendency to absorb Ag^+ ions from the solution. Thus end point will be reached earlier. Therefore, in order to avoid premature end point, the solution is shaken vigorously and constantly.

11.4. Adsorption Indicator

"Adsorption indicator is a substance which is adsorbed on the precipitate causing the production of a new colour."

In FaJan's method (1923-24), adsorption indicator is used in the titration of halide with silver nitrate solution. It does not give colour change in the solution, but gives the colour change on the surface of the precipitate only.

When silver nitrate solution is gradually added from a burette to a chloride solution (say NaCl solution), white precipitate of AgCl is produced. Since AgCl has a tendency to adsorb Cl^- ions, AgCl is surrounded by Cl^- ions present in the solution. Adsorbed Cl^- ions form a primary layer. This layer attracts sodium ions present in the solution. The adsorbed sodium ions form a secondary layer.

$$Ag^+ + Cl^- \rightarrow \underset{\text{(White ppt.)}}{AgCl}$$

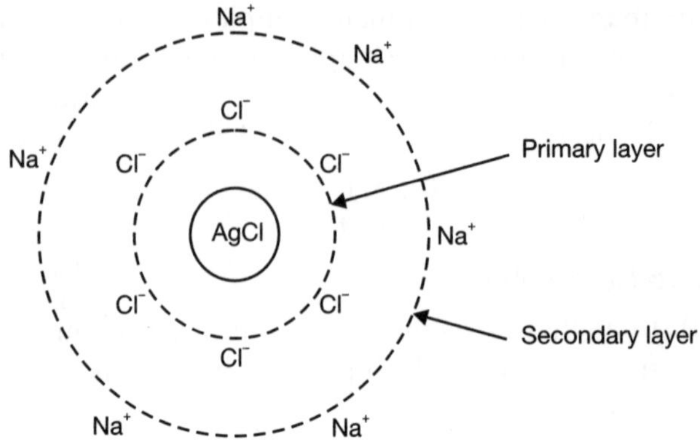

When equivalence point is reached and a slight excess amount of $AgNO_3$ solution is added, silver ions are adsorbed in primary layer and nitrate ions are adsorbed in secondary layer.

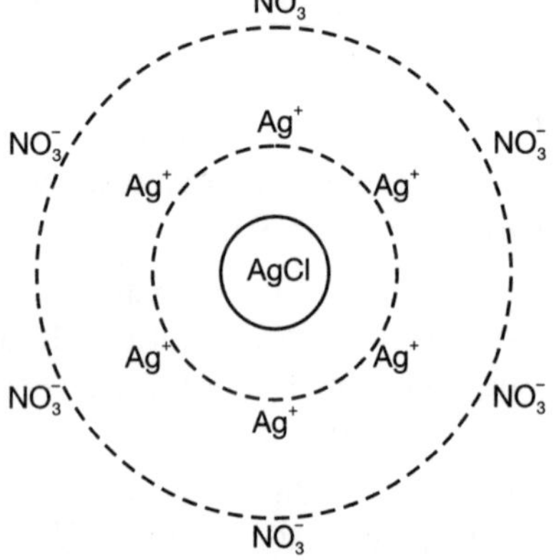

If adsorption indicator like fluorescein be present, then at equivalent point nitrate ions are replaced by fluorescein anions. It is so because nitrate ions are not strongly held by the particle while fluorescein anions are strongly held. As fluorescein anions get adsorbed, a pink or red colour appears.

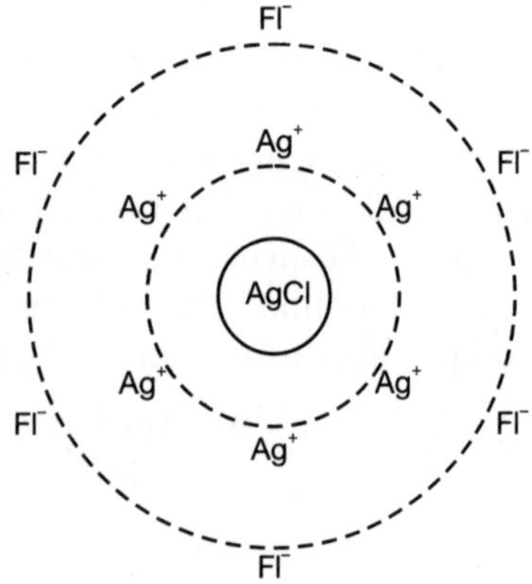

Fl^- indicates fluorescein anion.

Commonly used indicators for titrations of chloride against silver nitrate are:

(i) **Fluorescein:** It is a suitable indicator for very dilute solution of chloride with silver nitrate in neutral or faintly acidic solution. At the end point there is a sudden change of white precipitate in greenish-yellow medium to distinctly red.

(ii) **Dichlorofluorescein:** It is a suitable indicator for every dilute solution of chloride (say drinking water) and works even in the presence of acetic acid and weakly acid solution. The end point is from yellowish-green to red.

(iii) **Eosin:** It is employed for titration of bromide and iodide with silver nitrate (from burette) in the presence of acetic acid. The end point is the change from pink to reddish-violet. The halide solution should be diluted to 100 c.c.

(iv) **Di-iodo diethylfluorescein:** It is a better indicator for the determination of iodides. The solution of the iodide should be diluted to 100 c.c. before adding the indicator. It works in neutral or weakly acid medium. End point is the change of orange red to blue red when an iodide is titrated with silver nitrate (from burette).

11.5. Metal Ion Indicator

Metal ion indicator is used in complexometric titration, i.e., EDTA titration. Actually disodium salt of EDTA is used. The EDTA is represented as H_4Y. Hence disodium salt of EDTA is represented as Na_2H_2Y. Thus in short EDTA may be denoted as H_2Y^{2-}. In titration, the EDTA is taken in the burette and the solution of the metal ion is taken in the conical flask. When the EDTA is added into the conical flask, it forms stable complexes with metal ions. The reactions of the EDTA with metal ions are in the ratio of 1:1 regardless of the charge on the metal ions.

$$M^{2+} + \underset{\text{(EDTA)}}{H_2Y^{2-}} \rightarrow \underset{\text{(Complex)}}{MY^{2-}} + 2H^+$$

$$M^{3+} + \underset{\text{(EDTA)}}{H_2Y^{2-}} \rightarrow \underset{\text{(Complex)}}{MY^{-}} + 2H^+$$

$$\underset{}{M^{4+}} + \underset{\text{(EDTA)}}{H_2Y^{2-}} \rightarrow \underset{\text{(Complex)}}{MY} + 2H^+$$

When metal ion indicator (say EBT) is added into the conical flask containing metal ions, it forms a less stable complex with metal ion. In this stage, colour of the content of the conical flask remains red. At equivalent point, the indicator releases metal ions and the EDTA forms stable complexes with these metal ions. In this stage (end point), colour of the solution becomes blue.

The most widely used indicator is Eriochrome Black T (EBT). It is a complex organic chelating agent possessing the following formula:

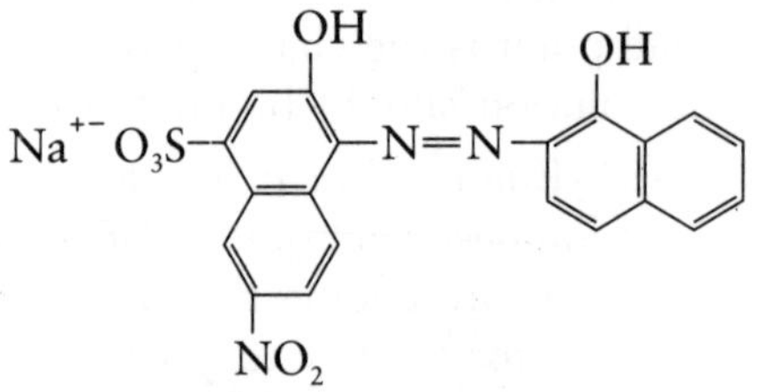

1–(1–Hydroxy–2–naphthylazo)–6–nitro–2–naphthol–4–sulphonate

The EBT is denoted as NaH_2In or H_2In^-. The phenolic group on EBT may also get ionised at high pH value (alkaline medium). Dissociation of indicator may be shown as under:

$$\underset{\text{(Red)}}{H_2In^-} + H_2O \rightarrow \underset{\text{(Blue)}}{HIn^{2-}} + H_3O^+$$

$$\underset{\text{(Blue)}}{HIn^{2-}} + H_2O \rightarrow \underset{\text{(Orange)}}{In^{3-}} + H_3O^+$$

Dissociation depends upon pH of the solution. DBT is a pH sensitive indicator.

$pH < 6$; the indicator exists as H_2In^- (red).

$pH > 7$; the indicator exists as HIn^{2-} (blue).

$pH > 10$; the indicator exists as In^{3-} (orange).

Thus, pH of the solution (of conical flask) should be maintained by suitable buffer solution (NH_4Cl + NH_4OH). The EDTA titration is performed in buffer of pH 8–10, the range in which indicator is blue (HIn^{2-}). As metal ions are present in the conical flask, indicator combines immediately with metal ions to form complexes. This causes the development of red colour. At the end point, the indicator will be set free because of the formation of stable metal-EDTA complex. This causes the development of blue colour.

$$\underset{\text{(Blue)}}{HIn^{2-}} + \underset{\text{(Metal ion)}}{M^{2+}} \rightarrow \underset{\text{(Red)}}{MIn^{-}} + H^{+}$$

$$\underset{\text{(Red)}}{MIn^{-}} + \underset{\text{(EDTA)}}{H_2Y^{2-}} \rightarrow \underset{\text{(Colourless)}}{MY^{2-}} + \underset{\text{(Blue)}}{HIn^{2-}} + H^{+}$$

12. APPARATUS USED IN VOLUMETRIC ANALYSIS

12.1. Chemical Balance

The instrument used for the determination of the weights of bodies is called balance. The chemical (analytical) balance differs from the ordinary balance in sensitiveness and accuracy. The essential features of a typical chemical balance are shown diagrammatically in Fig. 2 below:

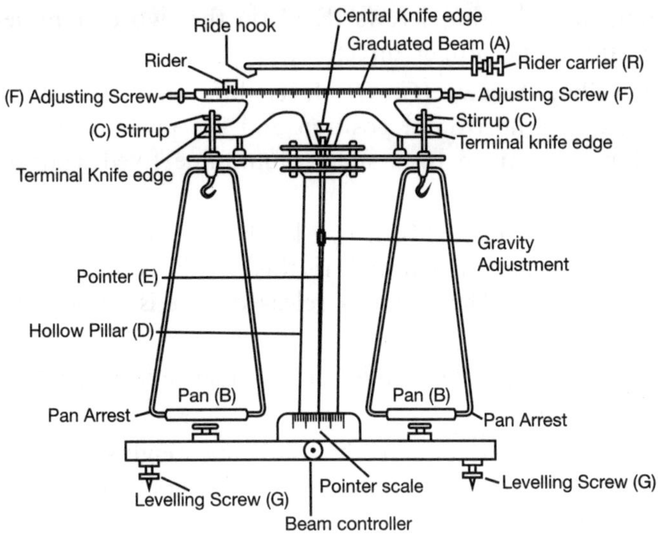

Fig. 2

12.1.1. Description of a chemical balance

Beam (A): It is a horizontal graduated metallic beam. The beam is provided with an agate knife edge at the centre. There are two other agate knife edges (pointing upwards) on the two extremities of the beam.

Pans (B, B): It is lighter metallic pan suspended by means of a stirrups (C, C) containing agate planes.

Hollow Pillar (D): It is hollow pillar fixed vertically to the base of the balance. The knife edge at the centre of the beam works upon an agate plane fixed to the top of a movable metal rod inside the hollow pillar.

Pointer (E): A long pointer is attached to the centre on the beam. This moves over a scale(s) at the foot of the pillar and serves to indicate the deflection of the beam from the horizontal position when the balance is in operation. When rightly adjusted, the pointer should move equal divisions on each side of this scale.

Adjusting Screws (F, F): Adjustment can be made by turning the adjusting screws (F, F) on the two extremities of the beam in the proper direction.

Levelling Screws (G, G): It enables the balance to be adjusted to a horizontal position and the pillar to a vertical position ascertained by a small plumb-line fixed against the pillar.

Beam Controller (H): It is a knob in front of the balance case. The beam is arrested if this knob is turned to its clockwise limit. The beam is released if it is turned slowly counter clockwise.

Rider Carrier (R): It slides on an independent metallic rod fixed horizontally.

Weight Box: It contains the following weights arragned as follows. The weights are made up of non-magnetic nickel-chromium alloy:

50 5	20 2	20 2	10 1	Gram
500 50	200 20	200 20	100 10	Miligram

12.1.2. How to use chemical balance

First of all the balance is levelled with the help of levelling screws and the plumb-line.

If the pointer does not move equal divisions on each side of the zero of the scale, then balance is adjusted with the help of adjusting screws.

Weights and the object to be weighed should always be placed in the centre of the pan.

Substance must be weighed in suitable containers, such as small beakers, weighing tubes, crucibles, or watch glasses.

Nothing should be added or removed from the pan when the knob is raised.

The balance must not be overloaded.

The knob of the balance should always be raised and lowered very gently and tried to avoid jerks as far as possible.

Nothing must be left on the pans when the weighing has been completed.

Balance should be closed before leaving the balance room.

12.2. Burette

It consists of a long graduated cylindrical glass tube with a uniform cross-section. It is provided with a glass stopcock at the lower end. Sometimes, the burette is fitted instead of having a glass tip, with a rubber tubing, a jet and a pinch cock. The burette is generally graduated from 0–50 c.c. Some instructions may be provided while using a burette:

- Before using a burette, it must be ascertained that it is not greasy. No drops should hang from the surface of glass after the liquid is run out from the burette.

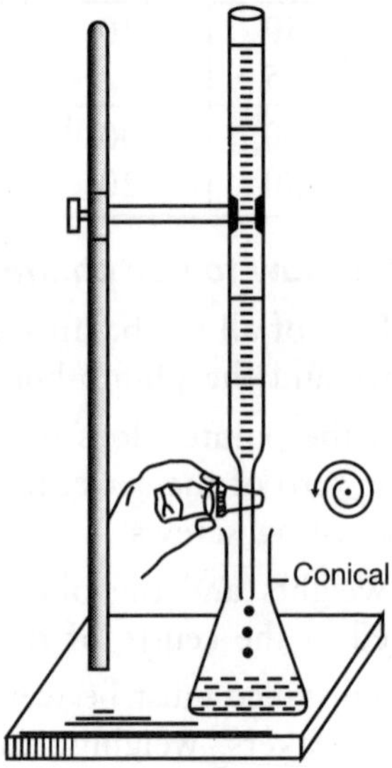

Fig. 3

- One should see that the glass tip of the burette works smoothly without leakage. If it does not, it must be lubricated with a little vaseline or grease.
- The burette must be supported in a stand in vertical position, by means of a clamp properly padded with cork or paper. It should never be held in a metal clamp directly, as it might break.
- The burette should be filled with the solution using a small burette funnel held in the left hand.
- Perhaps the best device for the accurate burette reading consists in holding behind the meniscus a piece of white paper.
- For all colourless solutions in the burette, lower meniscus should be noted and for coloured solutions like potassium permanganate or potassium dichromate, upper meniscus should be noted.

12.3. Pipette

This is an instrument which delivers a definite volume of a liquid. It consists of a glass tube with a bulb in the middle, on which is marked the volume of the pipette. The upper tube is etched around with a graduated mark while the lower tube is drawn into a fine jet.

Some instructions may be provided when using the pipette:

- The pipette is washed first with water then with a little solution to be sucked.

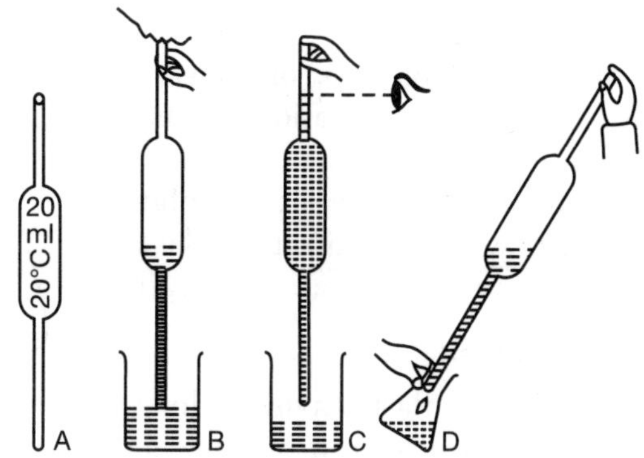

Fig. 4

- The pipette is inserted into the solution with the tip at or nearly touching the bottom of the container. By placing the upper end in the mouth, the solution is sucked gradually till the level of the solution rises well above the circular edged mark. Now the tube is removed from the mouth and the upper end closed by the forefinger (not by the thumb) immediately allowing the level of the solution to fall slowly but regularly until the lower part of the meniscus coincides with the circular mark.
- To discharge the pipette, the nozzle of it is introduced into the titration flask and the pressure released completely by removing the forefinger. The whole of the liquid will flow out automatically except a little of it in the nozzle. This should not be blown out but the nozzle should simply be touched with the liquid surface of the inner wall of the titration flask for about 10 seconds.
- Pipette should never be held at the bulb.
- A pipette with a broken nozzle should not be used.
- Poisonous solutions like arsenious oxide solution should not be sucked with a pipette.
- Hot liquids should not be measured with pipette.

12.4. Titration Flask

It is a special type of flask of convenient size used for carrying out titrations. It has a wide neck, a large belly and a flat bottom. The wide neck permits the free flow of the solution from the burette while the titration flask is shaken gradually during the titration operation.

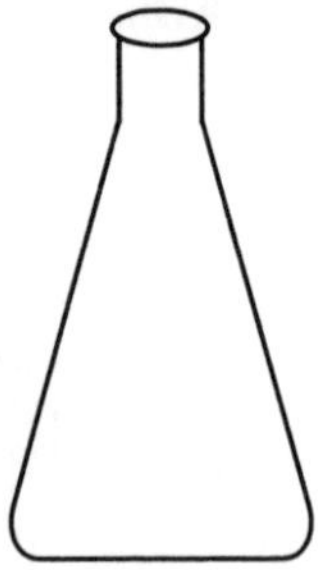

Fig. 5

12.5. Measuring Flask

It is a flat bottom flask having a long neck with a mark etched on it. It is provided with a ground glass stopper for shaking the solution when required. The required capacity of the flask is also etched on it. The most common sizes used in the laboratory are 100, 500, 1000 c.c. flasks. It is used for preparing standard solutions. The capacity of the flask at a definite temperature remains marked on the flask. A graduated flask when filled to mark contains a fixed volume of the liquid at a definite temperature.

Fig. 6

Precautions in the use of measuring flask are as follows:

- Flask should never be heated and hot liquids should not be placed in it.
- Its stopper should not be changed.
- It should not be rinsed with any solution. It is always washed with distilled water.
- Concentrated solutions of alkalis should not be prepared and stored in the measuring flask.

12.6. Weighing Tube

It is a flat bottomed cylindrical glass tube with a glass stopper. It is used for weighing the substances.

13. ACIDIMETRY AND ALKALIMETRY

13.1. To Prepare 0.1N HCl from the concentrated acid, when 0.1N Na_2CO_3 has been given

Theory: In order to prepare exactly 0.1N HCl, a stronger solution is first prepared. This solution is then titrated with 0.1N Na_2CO_3 solution using methyl orange as indicator. Thus normality of acid solution can be calculated. Then the acid solution can be accordingly diluted to get 0.1N HCl.

$$Na_2CO_3 + 2HCl \xrightarrow{M.O.} 2NaCl + H_2O + CO_2 \uparrow$$

Procedure: It involves the following steps:

(i) Making stronger solution: First of all 3 c.c. of concentrated hydrochloric acid is measured with a graduated cylinder or a burette and taken into a 250 c.c. measuring flask. Distilled water is added to it with shaking and the volume is made up to the mark. Stopper is fitted into the flask and shaken vigorously.

(ii) Titration: The burette is rinsed with hydrochloric acid solution (made above) and then filled by it. Now 25 c.c. of 0.1N Na_2CO_3 solution is pipetted out and taken into a conical flask. Two drops of methyl orange are added to it. The solution becomes yellow. Now hydrochloric acid solution is added dropwise from the burette. The addition of acid is continued until the solution becomes orange. The same process is repeated to get at least two concordant readings.

(iii) Dilution: From the above titration, normality of the acid is calculated. To make 100 c.c. 0.1N HCl solution, calculated amount of the acid solution is measured with a burette and taken into a 100 c.c. flask. Then distilled water is added to make its volume upto the mark. This solution formed is exactly 0.1N HCl.

Observation and Calculation:

Volume of concentrated HCl taken = 3 c.c.

Volume of HCl solution (dilute) containing 3 c.c. conc. HCl = 250 c.c.

Table 4

Sl. No.	Volume of 0.1N Na_2CO_3 taken in conical flask (c.c.)	Volume of HCl solution (dilute) taken in burette (c.c.)			Concordant reading (c.c.) (say)	Indicator	End point
		Initial reading (say)	Final reading (say)	Difference (say)			
1.	25	5.4	22.8	17.4		Methyl	
2.	25	22.8	40.1	17.3	17.3	orange	Orange
3.	25	3.2	20.5	17.3			

25 c.c. 0.1N Na_2CO_3 solution ≡ 17.3 c.c. HCl solution (dil.)

or, $V_1N_1 = V_2N_2$

or, $25 \times 0.1 = 17.3 \times N_2$

or, $N_2 = \dfrac{25 \times 0.1}{17.3}$

or, $N_2 = 0.145$

Let x c.c. of 0.145N HCl ≡ 100 c.c. 0.1N HCl

or, $V_1N_1 = V_2N_2$

or, $x \times 0.145 = 100 \times 0.1$

or, $x = \dfrac{100 \times 0.1}{0.145}$

or, $x = 68.97$ c.c.

By taking 68.97 c.c. of HCl solution (dilute), its volume is made 100 c.c. by addition of distilled water. The normality of the resulting solution is 0.1N.

13.2. To prepare 0.1N Na_2CO_3 solution and to find out the strength of the supplied Na_2CO_3 solution using hydrochloric acid as an intermediate solution

Theory: Molecular wt. of anhydrous $Na_2CO_3 = 106$

Equivalent wt. of anhydrous $Na_2CO_3 = \dfrac{106}{2} = 53$

∵ For N solution, 1000 c.c. solution should contain 53 g of Na_2CO_3.

$\therefore$ For $\frac{N}{10}$ solution, 1000 c.c. solution should contain $\frac{53}{10}$g of Na_2CO_3.

$\therefore$ For $\frac{N}{10}$ solution, 250 c.c. solution should contain $\frac{53 \times 250}{10 \times 1000}$g of Na_2CO_3 = 1.325 g of Na_2CO_3.

Thus 1.325 g of anhydrous Na_2CO_3 is dissolved in 250 c.c. of water to get 0.1N Na_2CO_3 solution. Then normality of hydrochloric acid solution is determined by titrating it with 0.1N Na_2CO_3 solution. At last, the strength of the supplied solution of sodium carbonate is determined by titrating it with given hydrochloric acid solution.

$$Na_2CO_3 + 2HCl \rightarrow 2NaCl + H_2O + CO_2 \uparrow$$

Procedure: It involves the following steps:

(i) Preparation of 0.1N Na_2CO_3 solution: First of all 1.325 g of A.R. quality anhydrous sodium carbonate is weighed and taken into 250 c.c. measuring flask. Distilled water is added in instalments and shaken to dissolve. At last, distilled water is added to make the volume up to the mark. Now this is 0.1N Na_2CO_3 solution.

(ii) Determination of strength of hydrochloric acid solution: The burette is rinsed with hydrochloric acid solution and filled with it. Initial reading of burette is noted. 25 c.c. of 0.1N Na_2CO_3 solution is pipetted out and taken into the conical flask. 2-3 drops of methyl orange are added to the conical flask. Now hydrochloric acid solution is added to it. When the orange colouration is developed in the flask, addition of hydrochloric acid is stopped and the final reading of the burette is noted. Now by calculation, normality of the acid solution is determined.

(iii) Determination of strength of supplied sodium carbonate solution: 25 c.c. of supplied sodium carbonate solution is taken in a conical flask. 2-3 drops of methyl orange are added to it. Now it is titrated with hydrochloric acid solution of known normality. Then by calculation, strength of the supplied solution of sodium carbonate is determined.

Observation and Calculation:

1.325 g of anhydrous sodium carbonate is dissolved in distilled water to obtain 250 c.c. 0.1N Na_2CO_3 solution.

Table 5

Sl. No.	Volume of 0.1N Na_2CO_3 (c.c.)	Volume of HCl taken in burette (c.c.)			Concordant reading (c.c.) (say)	Indicator	End point
		Initial reading (say)	Final reading (say)	Difference (say)			
1.	25	3.2	28.3	25.1			
2.	25	2.0	27.0	25.0	25.0	M.O.	Orange
3.	25	5.4	30.4	25.0			

25 c.c. HCl solution ≡ 25 c.c. 0.1N Na_2CO_3

or, $V_1N_1 = V_2N_2$

or, $25 \times N_1 = 25 \times 0.1$

or, $N_1 = \dfrac{25 \times 0.1}{25}$

or, $N_1 = 0.1$

or, Normality of HCl = 0.1N

Table 6

Sl. No.	Volume of supplied Na_2CO_3 solution (c.c.)	Volume of 0.1N HCl taken in burette (c.c.)			Concordant reading (c.c.) (say)	Indicator	End point
		Initial reading (say)	Final reading (say)	Difference (say)			
1.	25	0.0	24.1	24.1			
2.	25	24.1	48.1	24.0	24.0	M.O.	Orange
3.	25	0.0	24.0	24.0			

25 c.c. supplied Na_2CO_3 solution ≡ 24 c.c. 0.1N HCl solution

or, $V_1N_1 = V_2N_2$

or, $25 \times N_1 = 24 \times 0.1$

or, $N_1 = \dfrac{24 \times 0.1}{25}$

or, $N_1 = 0.096$

or, Normality of supplied Na_2CO_3 solution = 0.096N

∴ Gram/litre = Normality × Eq. wt.

= 0.096 × 53

= 5.088

∴ Strength of supplied Na_2CO_3 solution = 5.088 g/litre

13.3. To prepare 0.1N oxalic acid and to determine the strength of supplied sodium hydroxide solution

Theory: Molecular wt. of $H_2C_2O_4 \cdot 2H_2O$ = 126

Equivalent wt. of $H_2C_2O_4 \cdot 2H_2O = \dfrac{126}{2} = 63$

∵ For N solution, 1000 c.c. solution should contain 63 g of hydrated oxalic acid.

∴ For $\dfrac{N}{10}$ solution, 1000 c.c. solution should contain $\dfrac{63}{10}$ g of hydrated oxalic acid.

∴ For $\dfrac{N}{10}$ solution, 250 c.c. solution should contain $\dfrac{63 \times 250}{10 \times 1000}$ g of hydrated oxalic acid = 1.575 g of hydrated oxalic acid.

Thus 1.575 g of hydrated oxalic acid is dissolved in 250 c.c. of water to get 0.1N oxalic acid solution. Then supplied NaOH solution is titrated against 0.1N oxalic acid, using phenolphthalein as an indicator.

$$\begin{array}{l}COOH\\ |\\ COOH\end{array} + 2NaOH \xrightarrow{\text{Phenolphthalein}} \begin{array}{l}COONa\\ |\\ COONa\end{array} + 2H_2O$$

Procedure: It involves the following steps:

(i) Preparation of 0.1N oxalic acid: First of all 1.575 g of hydrated oxalic acid is weighed and taken into 250 c.c.

measuring flask. Distilled water is added to it and the flask is shaken to dissolve the acid. Now the volume of the solution is made up to the mark and the flask shaken well for homogenisation.

The solution so obtained is 0.1N oxalic acid.

(ii) **Determination of strength of supplied NaOH:** The burette is rinsed with 0.1N oxalic acid and then filled with the same. Initial reading of the burette is taken. 25 c.c. of the supplied NaOH solution is pipetted out and taken into the conical flask. Few drops of phenolphthalein are added to it. The solution becomes pink. Now the oxalic acid solution is run into the flask until the solution becomes colourless. The final reading of the burette is noted.

Observation and Calculation:

1.575 g of hydrated oxalic acid is dissolved in distilled water and the volume of the solution is made 250 c.c. The solution so obtained is now 0.1N oxalic acid.

Table 7

Sl. No.	Vol. of supplied NaOH (c.c.)	Volume of 0.1N oxalic acid taken in burette (c.c.)			Concordant reading (c.c.) (say)	Indicator	End point
		Initial reading (say)	Final reading (say)	Difference (say)			
1.	25	1.2	25.3	24.1			
2.	25	5.3	29.3	24.0	24.0	Phenol-	Colour-
3.	25	0.0	24.0	24.0		phthalein	less

25 c.c. NaOH solution = 24 c.c. 0.1N oxalic acid

or, $V_1N_1 = V_2N_2$

or, $25 \times N_1 = 24 \times 0.1$

or, $N_1 = \dfrac{24 \times 0.1}{25}$

or, $N_1 = 0.096$

∴ Gram/litre = Normality × Eq. wt.

	Mol. wt. of NaOH = 40
$= 0.096 \times 40$	Eq. wt. of NaOH $= \frac{40}{1}$
$= 3.84$	$= 40$

∴ Strength of NaOH = 3.84 g/litre

13.4. To estimate the strength of sodium carbonate and sodium bicarbonate present in a solution using 0.1N HCl

Theory: This determination is based on the selective use of indicators. When methyl orange is used as an indicator, the volume of 0.1N HCl used in the titration will neutralise both sodium carbonate and sodium bicarbonate completely. Suppose this volume of acid is x c.c.

$$Na_2CO_3 + 2HCl \xrightarrow{\text{M.O.}} 2NaCl + H_2O + CO_2$$

$$NaHCO_3 + HCl \xrightarrow{\text{M.O.}} NaCl + H_2O + CO_2$$

When phenolphthalein is used as an indicator, the volume of 0.1N HCl used in the titration will neutralise only half of the sodium carbonate. Suppose this volume of acid is y c.c.

$$Na_2CO_3 + HCl \xrightarrow{\text{Phenolphthalein}} NaHCO_3 + NaCl$$

∴ The vol. of acid for complete neutralisation of Na_2CO_3 = 2 y c.c.

∴ The vol. of acid for complete neutralisation of $NaHCO_3$ = (x − 2y) c.c.

Procedure: It involves the following steps:

(i) Titration using methyl orange: The burette is rinsed with 0.1N HCl and filled with the same. 25 c.c. of solution of alkali mixture is pipetted out and taken into conical flask. 2-3 drops of methyl orange are added to it. Now 0.1N HCl is run from the burette into the flask until the yellow colour of solution changes to orange. The process is repeated until two concordant readings are obtained. Suppose the volume of the acid used is x c.c.

(ii) Titration using phenolphthalein: 25 c.c. of solution of alkali mixture is pipetted out and taken into a conical flask. 2-3 drops of phenolphthalein are added to it. The solution becomes pink. Now 0.1N HCl is run from the burette into the flask until the solution becomes colourless. Suppose the volume of acid used is y c.c.

Observation and Calculation:

Table 8

Sl. No.	Vol. of solution of alkali mix (c.c.)	Volume of 0.1N HCl taken in burette (c.c.)			Concordant reading (x c.c.) (say)	Indicator	End point
		Initial reading (say)	Final reading (say)	Difference (say)			
1.	25	1.7	26.8	25.1			
2.	25	1.0	26.0	25.0	25.0	M.O.	Orange
3.	25	5.4	30.4	25.0			

Table 9

Sl. No.	Vol. of solution of alkali mix (c.c.)	Volume of 0.1N HCl taken in burette (c.c.)			Concordant reading (y c.c.) (say)	Indicator	End point
		Initial reading (say)	Final reading (say)	Difference (say)			
1.	25	2.6	7.7	5.1			
2.	25	7.7	12.7	5.0	5.0	Phenolphthalein	Colourless
3.	25	12.7	17.7	5.0			

Vol. of acid to neutralise Na_2CO_3 = 2y c.c.
= 2 × 5 c.c.
= 10 c.c.

Vol. of acid to neutralise $NaHCO_3$ = (x – 2y) c.c.
= (25 – 10) c.c.
= 15 c.c.

(i) Estimation of Na_2CO_3:

25 c.c. Na_2CO_3 solution ≡ 10 c.c. 0.1N HCl

or, $V_1N_1 = V_2N_2$

or, $25 \times N_1 = 10 \times 0.1$

or, $N_1 = \frac{10 \times 0.1}{25}$

or, $N_1 = 0.04$

$\therefore$ Gram/litre = Normality × Eq. wt.

	Mol. wt. of Na_2CO_3 = 106
$= 0.04 \times 53$	Eq. wt. of Na_2CO_3 $= \frac{106}{2}$
$= 2.12$	$= 53$

$\therefore$ Strength of Na_2CO_3 = 2.12 g/litre

(ii) Estimation of $NaHCO_3$:

25 c.c. $NaHCO_3$ solution ≡ 15 c.c. 0.1N HCl

or, $V_1N_1 = V_2N_2$

or, $25 \times N_1 = 15 \times 0.1$

or, $N_1 = \frac{15 \times 0.1}{25}$

or, $N_1 = 0.06$

$\therefore$ Gram/litre = Normality × Eq. wt.

	Mol. wt. of $NaHCO_3$ = 84
$= 0.06 \times 84$	Eq. wt. of $NaHCO_3$ $= \frac{84}{1}$
$= 5.04$	$= 84$

$\therefore$ Strength of $NaHCO_3$ = 5.04 g/litre

Things to Remember

An alternative method for analysis of Na_2CO_3 and $NaHCO_3$ in the solution of their mixture can be proposed as under:

The solution of Na_2CO_3 and $NaHCO_3$ is titrated with 0.1N HCl using phenolphthalein as an indicator. The volume (say x c.c.) of the acid required to attain the end point reveals the amount required to neutralise half of the sodium carbonate.

$$Na_2CO_3 + HCl \xrightarrow{\text{Phenolphthalein}} NaHCO_3 + NaCl$$

Now the same solution of the conical flask is titrated with 0.1N HCl using methyl orange as an indicator. The volume (say y c.c.) of the acid required to attain the end point reveals the amount required to neutralise whole of the $NaHCO_3$ and remaining half of the Na_2CO_3 (which is present now as $NaHCO_3$).

$$NaHCO_3 + HCl \xrightarrow{M.O.} NaCl + H_2O + CO_2$$

$\therefore$ Vol. of acid to neutralise the whole Na_2CO_3 = 2x c.c.

$\therefore$ Vol. of acid to neutralise the whole $NaHCO_3$ = (y − x) c.c.

13.5. To determine the amounts of sodium carbonate and sodium hydroxide present together in the given solution using 0.1N HCl

Theory: If the mixture containing 2-3 drops of phenolphthalein is titrated against 0.1N HCl, the pink colour of the indicator gets decolourised when whole of the sodium hydroxide and half of the sodium carbonate get neutralised. Let volume of 0.1N HCl required for this work is x c.c.

$$NaOH + HCl \xrightarrow{\text{Phenolphthalein}} NaCl + H_2O$$

$$Na_2CO_3 + HCl \xrightarrow{\text{Phenolphthalein}} NaHCO_3 + NaCl$$

To the above neutralised solution, 2-3 drops of methyl orange are added and further titrated against 0.1N HCl. At end point the yellow colour changes to orange. Let the volume of 0.1N HCl required here is y c.c. This volume of acid corresponds to the quantity required to neutralise remaining half of the carbonate.

$$NaHCO_3 + HCl \xrightarrow{M.O.} NaCl + H_2O + CO_2$$

$\therefore$ Volume of 0.1N HCl required to neutralise Na_2CO_3 = 2y c.c.

$\therefore$ Volume of 0.1N HCl required to neutralise NaOH = (x − y) c.c.

Procedure: The burette is rinsed with 0.1N HCl and filled with the same. Initial reading is taken. 25 c.c. of solution of alkali mixture is pipetted out and taken into a conical flask. 2-3 drops of phenolphthalein are added to it. Now 0.1N HCl is run from burette into conical flask. At end point the pink colour of the solution becomes colourless. Final reading of the burette is noted. Then 2-3 drops of methyl orange are added to the same solution contained in the conical flask. The final reading of the first titration is considered as initial reading of the second titration. Now 0.1N HCl is run from burette into the flask. At end point the yellow colour of the solution changes to orange. Final reading of the burette is noted. The above process is repeated until two concordant readings are obtained.

Observation and Calculation:

Table 10

Vol. of alkali mix. (c.c.)	Vol. of 0.1N HCl taken in burette (Phenolphthalein) $NaOH + HCl \rightarrow NaCl + H_2O$ $Na_2CO_3 + HCl \rightarrow NaHCO_3 + NaCl$ (x c.c.)				Vol. of 0.1N HCl taken in burette (Methyl orange) $NaHCO_3 + HCl \rightarrow NaCl + H_2O + CO_2$ (y c.c.)			
	I.R.	F.R.	Diff.	C.R.	I.R.	F.R.	Diff.	C.R.
25	0.0	19.1	19.1		19.1	25.2	6.1	
25	2.5	21.5	19.0	19.0	21.5	27.5	6.0	6.0
25	4.2	23.2	19.0		23.2	29.2	6.0	

$\therefore$ Vol. of acid to neutralise Na_2CO_3 = 2y c.c.
= 2 × 6 c.c.
= 12 c.c.

$\therefore$ Vol. of acid to neutralise NaOH = (x − y) c.c.
= (19 − 6) c.c.
= 13 c.c.

(i) Estimation of Na_2CO_3:

25 c.c. Na_2CO_3 solution ≡ 12 c.c. 0.1N HCl

or, $V_1N_1 = V_2N_2$

or, $25 \times N_1 = 12 \times 0.1$

or, $N_1 = \frac{12 \times 0.1}{25}$

or, $N_1 = 0.048$

$\therefore$ Gram/litre = Normality × Eq. wt.
= 0.048 × 53
= 2.544

Mol. wt. of Na_2CO_3 = 106
Eq. wt. of Na_2CO_3 = 53

$\therefore$ Strength of Na_2CO_3 = 2.544 g/litre

(ii) Estimation of NaOH:

25 c.c. NaOH solution ≡ 13 c.c. 0.1N HCl

or, $V_1N_1 = V_2N_2$

or, $25 \times N_1 = 13 \times 0.1$

or, $N_1 = \dfrac{13 \times 0.1}{25}$

or, $N_1 = 0.052$

∴ Gram/litre = Normality × Eq. wt.

= 0.052 × 40 | Mol. wt. of NaOH = 40

= 2.08 | Eq. wt. of NaOH = 40

∴ Strength of NaOH = 2.08 g/litre

Things to Remember

Phenolphthalein should not be used as an indicator in the titration of Na_2CO_3 with HCl. It can be explained as under:

Na_2CO_3 is neutralised by HCl in two steps:

$$\underset{pH = 11}{Na_2CO_3} \xrightarrow{HCl} \underset{pH = 8.3}{NaHCO_3} \xrightarrow{HCl} \underset{pH = 3.8}{NaCl}$$

Na_2CO_3 is strongly alkaline. Its pH value is greater than 7 (say 11). When HCl is added to this solution, it is converted into sodium bicarbonate.

$$Na_2CO_3 + HCl \rightarrow NaHCO_3 + NaCl \qquad ... (1)$$

When total Na_2CO_3 is converted into $NaHCO_3$, the pH of the resulting solution becomes 8.3. Phenolphthalein has its pH range 8 to 10. So it acts as an indicator for the above reaction (1). When more HCl is added, $NaHCO_3$ is converted into NaCl.

$$NaHCO_3 + HCl \rightarrow NaCl + H_2O + CO_2 \qquad ... (2)$$

At the equivalence point of this reaction, the pH of the solution becomes 3.8. In this titration, fall in pH is 8.3 to 3.8. Since phenolphthalein changes its colour over the pH range 8 to 10, hence it does not act as an indicator for reaction (2).

Hence phenolphthalein is not used as an indicator in the titration of Na_2CO_3 with HCl.

13.6. To determine the amounts of sodium hydroxide and sodium bicarbonate present together in the given solution using 0.1N HCl

Theory: If the mixture containing phenolphthalein as the indicator is titrated against 0.1N HCl, the pink colour of the solution gets discharged when whole of sodium hydroxide gets neutralised. Let the vol. of the acid required is x c.c.

$$NaOH + HCl \xrightarrow{\text{Phenolphthalein}} NaCl + H_2O$$

To the above neutralised solution, 2-3 drops of methyl orange are added and further 0.1N HCl is added from burette till orange colour is obtained. This quantity of acid added corresponds to the quantity required to neutralise whole of the bicarbonate. Let the volume of the acid required in second titration is y c.c.

$$NaHCO_3 + HCl \xrightarrow{\text{M.O.}} NaCl + H_2O + CO_2$$

Procedure: The burette is rinsed with 0.1N HCl and filled with the same. Initial reading is taken. 25 c.c. of the solution of the alkali mixture is pipetted out and taken into the conical flask. Now 2-3 drops of phenolphthalein are added to it and titrated against 0.1N HCl. At the end point pink colour of the solution becomes colourless. Final reading of the burette is noted. Now in the same solution 2-3 drops of methyl orange are added and further 0.1N HCl is added. At the end point yellow colour of the solution turns into orange. Final reading of the burette is noted. The final reading of the titration will of course be initial reading of the second titration. The process is repeated till two concordant readings are obtained.

Observation and Calculation:

Table 11

Vol. of mix. (c.c.)	Vol. of 0.1N HCl taken in burette (Phenolphthalein) $NaOH + HCl \rightarrow NaCl + H_2O$ (x c.c.)				Vol. of 0.1N HCl taken in burette (M.O.) $NaHCO_3 + HCl \rightarrow NaCl + H_2O + CO_2$ (y c.c.)			
	I.R.	F.R.	Diff.	C.R.	I.R.	F.R.	Diff.	C.R.
25	1.4	25.5	24.1		24.1	48.6	24.5	
25	0.0	24.0	24.0	24.0	24.0	48.4	24.4	24.4
25	0.4	24.4	24.0		24.4	48.8	24.4	

∴ Vol. of acid required to neutralise NaOH = 24.0 c.c.

∴ Vol. of acid required to neutralise $NaHCO_3$ = 24.4 c.c.

(i) Estimation of NaOH:

25 c.c. NaOH solution ≡ 24 c.c. 0.1N HCl

or, $V_1N_1 = V_2N_2$

or, $25 \times N_1 = 24 \times 0.1$

or, $N_1 = \dfrac{24 \times 0.1}{25}$

or, $N_1 = 0.096$

∴ Gram/litre = Normality × Eq. wt.

= 0.096 × 40 | Mol. wt. of NaOH = 40

= 3.84 | Eq. wt. of NaOH = 40

∴ Strength of NaOH = 3.84 g/litre

(ii) Estimation of $NaHCO_3$:

25 c.c. $NaHCO_3$ solution ≡ 24.4 c.c. 0.1N HCl

or, $V_1N_1 = V_2N_2$

or, $25 \times N_1 = 24.4 \times 0.1$

or, $N_1 = \dfrac{24.4 \times 0.1}{25}$

or, $N_1 = 0.0976$

∴ Gram/litre

= Normality × Eq. wt.

= 0.0976 × 84 | Mol. wt. of $NaHCO_3$ = 84

= 8.198 | Eq. wt. of $NaHCO_3$ = 84

∴ Strength of $NaHCO_3$ = 8.198 g/litre

14. REDOX TITRATION

14.1. To prepare 0.1N $KMnO_4$ solution and to standardise it using 0.1N oxalic acid

Theory: $MnO_4^- + 8H^+ + 5e \rightarrow Mn^{2+} + 4H_2O$

∴ Eq. wt. of $KMnO_4$ in acidic medium

$$= \frac{\text{Molecular wt. of } KMnO_4}{\text{Number of electrons taken}}$$

$$= \frac{39 + 55 + 64}{5}$$

$$= \frac{158}{5}$$

$$= 31.6$$

∵ For 1000 c.c. N solution, 31.6 g $KMnO_4$ is needed.

∴ For 1000 c.c. 0.1N solution, 31.6 × 0.1 g $KMnO_4$ is needed = 3.16 g $KMnO_4$.

Thus 3.16 g potassium permanganate is dissolved in distilled water and the volume of the solution is made 1000 c.c. This solution is roughly 0.1N $KMnO_4$. To standardise this solution it is titrated against 0.1N oxalic acid and by calculation factor is assigned to it so that exact normality of the solution can be known.

$$\left[MnO_4^- + 8H^+ + \cancel{5e} \rightarrow Mn^{2+} + 4H_2O\right] \times 2$$

$$\left[C_2O_4^{2-} \rightarrow 2CO_2 + \cancel{2e}\right] \times 5$$

$$\underset{\text{(Violet)}}{2MnO_4^-} + 5C_2O_4^{2-} + 16H^+ \rightarrow \underset{\text{(Colourless)}}{2Mn^{2+}} + 8H_2O + 10CO_2$$

Procedure: It involves following steps:

(i) Preparation of roughly 0.1N $KMnO_4$: First of all 3.16 g potassium permanganate (A.R.) is weighed and taken into 500 c.c. beaker containing 100 c.c. distilled water. The solution is kept for 2-3 days or heated to boiling for sometime by keeping the beaker on a water bath. It is cooled to room temperature and filtered through glass wool into a 1000 c.c. measuring flask. The volume of the solution is made up to the mark, i.e., 1000 c.c. This solution is now roughly 0.1N $KMnO_4$.

(ii) Standardisation: The burette is rinsed with $KMnO_4$ solution and filled with the same. 25 c.c. of 0.1N oxalic acid is pipetted out and taken into a conical flask. One test tube of dilute sulphuric acid is added to it. The conical flask is heated on a wire gauze to 60–70°C, i.e., to a temperature which is just unbearable to touch by hand. Now initial reading of the burette is noted and potassium permanganate solution is run from burette into the conical flask. The conical flask is stirred vigorously and constantly during the addition of permanganate solution. At the end point the solution of the conical flask becomes pink. Then final reading of the burette is noted. This titration is repeated till two concordant readings are obtained.

Observation and Calculation:

Table 12

<table>
<tr><th rowspan="2">Sl. No.</th><th rowspan="2">Volume of 0.1N oxalic acid (c.c.)</th><th colspan="3">Volume of roughly 0.1N $KMnO_4$ taken in burette (c.c.)</th><th rowspan="2">Concordant reading (c.c.) (say)</th><th rowspan="2">Indicator</th><th rowspan="2">End point</th></tr>
<tr><th>Initial reading (say)</th><th>Final reading (say)</th><th>Diff. (say)</th></tr>
<tr><td>1.</td><td>25</td><td>5.5</td><td>30.9</td><td>25.4</td><td rowspan="3">25.3</td><td rowspan="3">Self indicator</td><td rowspan="3">Pink</td></tr>
<tr><td>2.</td><td>25</td><td>4.2</td><td>29.5</td><td>25.3</td></tr>
<tr><td>3.</td><td>25</td><td>4.0</td><td>29.3</td><td>25.3</td></tr>
</table>

25.3 c.c. 0.1N (f = x) $KMnO_4 \equiv$ 25 c.c. 0.1N Oxalic acid

or, $V_1N_1 = V_2N_2$

or, $25.3 \times 0.1 \times x = 25 \times 0.1$

or, $x = \dfrac{25 \times 0.1}{25.3 \times 0.1}$

or, $x = 0.988$

$\therefore$ Actual strength of $KMnO_4$ = 0.1N (f = 0.988)

Things to Remember

(i) The titration which involves the use of potassium permanganate solution is known as 'Permanganometry'.

(ii) $KMnO_4$ is not a primary standard because:

- It cannot be obtained in pure state.
- On standing, potassium permanganate undergoes decomposition to MnO_2 in presence of impurities and light.

$$4MnO_4^- + 2H_2O \rightarrow 4MnO_2 + 4OH^- + 3O_2$$

(iii) Since $KMnO_4$ is not a primary standard, preparation of its standard solution must be followed by standardisation. Potassium permanganate solution may be standardised by titrating it with standard solution of either oxalic acid or Mohr's salt as they are primary standards.

(iv) A solution of potassium permanganate, once properly prepared is quite stable as long as it is completely protected from dust, excess of reducing gases, and direct sunlight or other strong light sources.

(v) During the titration one should be ensured that there is no air bubble in the nozzle of the burette.

(vi) Permanganometry needs no indicator. Potassium permanganate acts as self indicator.

(vii) $KMnO_4$ is only acidified by H_2SO_4. It should never be acidified by HCl or HNO_3. It is so because HCl is oxidised by $KMnO_4$ and HNO_3 may oxidise the titrand.

14.2. To prepare 0.1N oxalic acid and to estimate the strength of the given sample of $KMnO_4$ solution

Theory: Molecular wt. of $H_2C_2O_4{\cdot}2H_2O = 126$

$$\text{Eq. wt. of } H_2C_2O_4{\cdot}2H_2O = \frac{126}{2} = 63$$

∵ For 1000 c.c. N-solution, 63 g of hydrated oxalic acid is needed.

∴ For 1000 c.c. 0.1N-solution, 63×0.1 g of hydrated oxalic acid is needed = 6.3 g.

Thus to prepare 0.1N oxalic acid solution, 6.3 g of hydrated acid is dissolved in water and the volume is made 1000 c.c. This solution is then titrated against given potassium permanganate solution.

$$\left[MnO_4^- + 8H^+ + 5e \rightarrow Mn^{2+} + 4H_2O\right] \times 2$$

$$\left[C_2O_4^{2-} \rightarrow 2CO_2 + 2e\right] \times 5$$

$$2MnO_4^- + 5C_2O_4^{2-} + 16H^+ \rightarrow 2Mn^{2+} + 8H_2O + 10CO_2$$

Procedure: It involves following steps:

(i) **Preparation of 0.1N oxalic acid:** First of all 6.3 g of hydrated oxalic acid is weighed and taken into a 1000 c.c. measuring flask. Distilled water is added and shaken to dissolve the acid. Finally the volume of the solution is made up to the mark. Now 0.1N oxalic acid solution is ready.

(ii) **Titration:** The burette is rinsed with the given $KMnO_4$ solution and filled with the same. 25 c.c. of 0.1N oxalic acid is pipetted out and taken into a conical flask. One test tube of dilute sulphuric acid is added to it. The conical flask is heated on a wire gauze to 60–70°C, i.e., to a temperature which is just unbearable to touch by hand. Now the initial reading of the burette is noted and potassium permanganate solution is run from burette into the conical flask. The conical flask is stirred vigorously and constantly during the addition of permanganate solution. At the end point the solution of the conical flask becomes pink. Then final reading of the burette is noted. This titration is repeated till two concordant readings are obtained.

Observation and Calculation:

Table 13

Sl. No.	Volume of 0.1N oxalic acid (c.c.)	Volume of $KMnO_4$ solution taken in burette (c.c.)			Concordant reading (c.c.) (say)	Indicator	End point
		Initial reading (say)	Final reading (say)	Diff. (say)			
1.	25	0.0	25.6	25.6			
2.	25	2.2	27.7	25.5	25.5	Self indicator	Pink
3.	25	5.0	30.5	25.5			

25.5 c.c. $KMnO_4$ solution ≡ 25 c.c. 0.1N oxalic acid

or, $V_1N_1 = V_2N_2$

or, $25.5 \times N_1 = 25 \times 0.1$

or, $N_1 = \frac{25 \times 0.1}{25.5}$

or, $N_1 = 0.098$

$\therefore$ Gram/litre

$= \text{Normality} \times \text{Eq. wt.}$

$= 0.098 \times 31.6$ | Mol. wt. of $KMnO_4$ = 158

$= 3.097$ | Eq. wt. of $KMnO_4$ = 31.6

$\therefore$ Strength of $KMnO_4$ = 3.097 g/litre

14.3. To prepare 0.1N solution of Mohr's salt and to estimate the strength of given $KMnO_4$ solution

Theory: Mol. wt. of $FeSO_4 \cdot (NH_4)_2SO_4 \cdot 6H_2O$ = 392

Eq. wt. of $FeSO_4 \cdot (NH_4)_2SO_4 \cdot 6H_2O$ = 392

$\because$ For 1000 c.c. of N solution, 392 g of Mohr's salt is required.

$\therefore$ For 1000 c.c. of 0.1N solution, 392×0.1 g of Mohr's salt is required.

$\therefore$ For 250 c.c. of 0.1N solution, $\frac{392 \times 0.1 \times 250}{1000}$ g of Mohr's salt is required

= 9.8 g of Mohr's salt is required.

Thus 9.8 g of Mohr's salt is dissolved in distilled water and the volume is made 250 c.c. Now 0.1N Mohr's salt solution is ready. Then this solution is titrated against given $KMnO_4$ solution.

$$MnO_4^- + 8H^+ + \cancel{5e} \rightarrow Mn^{2+} + 4H_2O$$

$$[Fe^{2+} \rightarrow Fe^{3+} + \cancel{e}] \times 5$$

$$\overline{MnO_4^- + 5Fe^{2+} + 8H^+ \rightarrow Mn^{2+} + 5Fe^{3+} + 4H_2O}$$

Procedure: It involves following steps:

(i) Preparation of 0.1N Mohr's salt: First of all 9.8 g of Mohr's salt (A.R.) is weighed and taken into a 250 c.c. measuring flask which is already having a few drops of concentrated sulphuric acid. Then 50 c.c. distilled water is added and shaken to dissolve the Mohr's salt. Now the volume is made up to the mark by adding distilled water. Thus 0.1N Mohr's salt solution is ready.

(ii) Titration: The burette is rinsed with $KMnO_4$ solution and filled with the same. Initial reading of the burette is noted. Now 25 c.c. of 0.1N Mohr's salt solution is pipetted out and taken into a conical flask. One test tube of dilute sulphuric acid is added to it. Now $KMnO_4$ solution is run from the burette into the conical flask. The conical flask is shaken constantly during the addition of $KMnO_4$ solution. At the end point, the solution becomes pink. The final reading of the burette is noted. The titration is repeated till two concordant readings are obtained.

Observation and Calculation:

Table 14

Sl. No.	Vol. of 0.1N Mohr's salt (c.c.)	Volume of $KMnO_4$ taken in burette (c.c.)			Concordant reading (c.c.) (say)	Indicator	End point
		Initial reading (say)	Final reading (say)	Diff. (say)			
1.	25	0.2	25.0	24.8			
2.	25	25.0	49.7	24.7	24.7	Self indicator	Pink
3.	25	0.0	24.7	24.7			

24.7 c.c. $KMnO_4 \equiv$ 25 c.c. 0.1N Mohr's salt

or, $V_1N_1 = V_2N_2$

or, $24.7 \times N_1 = 25 \times 0.1$

or, $N_1 = \dfrac{25 \times 0.1}{24.7}$

or, $N_1 = 0.101$

$\therefore$ Gram/litre

= Normality × Eq. wt.

= 0.101×31.6 | Mol. wt. of $KMnO_4$ = 158

= 3.19 | Eq. wt. of $KMnO_4$ = 31.6

$\therefore$ Strength of $KMnO_4$ = 3.19 g/litre

Things to Remember

The test solution should contain dilute sulphuric acid to reduce the tendency for atmospheric oxidation. Presence of dilute H_2SO_4 also provides acidic medium in which potassium permanganate acts as strong oxidising agent.

14.4. To assign the factor to roughly decinormal solution of $KMnO_4$ with the help of 0.1N oxalic acid and to estimate the strength of given $NaNO_2$ solution using the $KMnO_4$ solution

Theory: 0.1N oxalic acid is titrated against the approximate 0.1N $KMnO_4$ solution and then by calculation factor is assigned to the roughly 0.1N $KMnO_4$.

$$\left[MnO_4^- + 8H^+ + \cancel{5e} \rightarrow Mn^{2+} + 4H_2O\right] \times 2$$

$$\left[C_2O_4^{2-} \rightarrow 2CO_2 + \cancel{2e}\right] \times 5$$

$$\overline{2MnO_4^- + 5C_2O_4^{2-} + 16H^+ \rightarrow 2Mn^{2+} + 8H_2O + 10CO_2}$$

Now the given sodium nitrite solution is titrated against 0.1N (f = x) $KMnO_4$ and then by calculation strength of sodium nitrite is calculated.

$$\left[MnO_4^- + 8H^+ + \cancel{5e} \rightarrow Mn^{2+} + 4H_2O\right] \times 2$$

$$\left[NO_2^- + H_2O \rightarrow NO_3^- + 2H^+ + \cancel{2e}\right] \times 5$$

$$\overline{2MnO_4^- + 5NO_2^- + 6H^+ \rightarrow 2Mn^{2+} + 5NO_3^- + 3H_2O}$$

Procedure: It involves following steps:

(i) To assign factor to roughly 0.1N $KMnO_4$: First of all the burette is rinsed with the $KMnO_4$ solution and filled with the same. Initial reading of the burette is noted. Now 25 c.c. of 0.1N oxalic acid is pipetted out and taken into a conical flask. One test tube of dilute sulphuric acid is added to it. The conical flask is heated on a wire gauze to 60–70°C. Then $KMnO_4$ solution is run from the burette into the conical flask. The conical flask is shaken constantly during the addition of the $KMnO_4$ solution. At the end point, colour of the solution becomes pink. The titration is repeated till two concordant readings are obtained.

(ii) Estimation of $NaNO_2$: The burette is rinsed with $NaNO_2$ solution and filled with the same. Initial reading of the burette is noted. Now 25 c.c. of 0.1N (f = x) $KMnO_4$ solution is pipetted out and taken into a conical flask. One test tube of dilute sulphuric acid is added to it. Then $NaNO_2$ solution is run from the burette into the conical flask. The conical flask is shaken constantly during the addition of the $NaNO_2$ solution. At the end point, the solution of the conical flask becomes colourless. The final reading of the burette is noted. The titration is repeated till two concordant readings are obtained.

Observation and Calculaticn:

Table 15

Sl. No.	Volume of 0.1N oxalic acid (c.c.)	Volume of 0.1N (f = x) $KMnO_4$ taken in burette (c.c.)			Concordant reading (c.c.) (say)	Indicator	End point
		Initial reading (say)	Final reading (say)	Diff. (say)			
1.	25	2.8	26.9	24.1			
2.	25	3.9	27.6	23.7	23.7	Self indicator	Pink
3.	25	1.2	24.9	23.7			

23.7 c.c. 0.1N (f = x) $KMnO_4 \equiv$ 25 c.c. 0.1N oxalic acid

or, $V_1N_1 = V_2N_2$

or, $23.7 \times 0.1 \times x = 25 \times 0.1$

or, $x = \dfrac{25 \times 0.1}{23.7 \times 0.1}$

or, $x = 1.05$

Thus the given $KMnO_4$ solution is 0.1N (f = 1.05)

Table 16

Sl. No.	Volume of 0.1N (f = 1.05) $KMnO_4$ (c.c.)	Volume of $NaNO_2$ taken in burette (c.c.)			Concordant reading (c.c.) (say)	Indicator	End point
		Initial reading (say)	Final reading (say)	Diff. (say)			
1.	25	1.35	27.65	26.30			
2.	25	1.55	27.80	26.25	26.25	Self indicator	Colour-less
3.	25	1.40	27.65	26.25			

26.25 c.c. $NaNO_2 \equiv 25$ c.c. $0.1N$ (f = 1.05) $KMnO_4$

or, $V_1N_1 = V_2N_2$

or, $26.25 \times N_1 = 25 \times 0.1 \times 1.05$

or, $N_1 = \dfrac{25 \times 0.1 \times 1.05}{26.25}$

or, $N_1 = 0.1$

$\therefore$ Gram/litre

= Normality × Eq. wt.

$= 0.1 \times 34.5$

$= 3.45$

$NO_2^- + H_2O \rightarrow NO_3^- + 2H^+ + 2e$

$\therefore$ Eq. wt. of $NaNO_2 = \dfrac{\text{Mol. wt.}}{2} = \dfrac{69}{2} = 34.5$

$\therefore$ Strength of $NaNO_2$ = 3.45 g/litre

Things to Remember

For estimation of $NaNO_2$, it is taken in the burette. If $NaNO_2$ solution is taken in the conical flask and dilute H_2SO_4 is added to it, HNO_2 is liberated.

$$2NaNO_2 + H_2SO_4 \rightarrow Na_2SO_4 + 2HNO_2$$

The nitrous acid (HNO_2) so produced is unstable and undergoes auto-oxidation.

$$3HNO_2 \rightarrow HNO_3 + H_2O + 2NO$$

Thus $NaNO_2$ solution should always be taken in the burette while titrating with $KMnO_4$ solution.

14.5. To estimate the amount of sulphuric acid and oxalic acid in the supplied solution of their mixture using 0.1N $KMnO_4$ and 0.1N NaOH

Theory: First of all the given acid mixture is titrated with 0.1N NaOH which reveals the total amount of the acids present in the mixture.

$$H_2SO_4 + 2NaOH \rightarrow Na_2SO_4 + 2H_2O$$

$$\begin{array}{l} COOH \\ | \\ COOH \end{array} + 2NaOH \rightarrow \begin{array}{l} COONa \\ | \\ COONa \end{array} + 2H_2O$$

In the second titration, the given acid mixture is titrated with 0.1N $KMnO_4$ which reveals the amount of oxalic acid present in the mixture.

$$\left[MnO_4^- + 8H^+ + \cancel{5e} \rightarrow Mn^{2+} + 4H_2O\right] \times 2$$

$$\left[C_2O_4^{2-} \rightarrow 2CO_2 + \cancel{2e}\right] \times 5$$

$$2MnO_4^- + 5C_2O_4^{2-} + 16H^+ \rightarrow 2Mn^{2+} + 8H_2O + 10CO_2$$

Procedure: It involves following steps:

(i) **Titration with 0.1N NaOH:** First of all the burette is rinsed with the acid mixture and filled with the same. Initial reading of the burette is noted. Now 25 c.c. of 0.1N NaOH is pipetted out and taken into the conical flask. 2-3 drops of phenolphthalein are added to it. Acid mixture is run from burette into the conical flask. At the end point, the solution of the conical flask becomes colourless. The final reading of the burette is noted.

(ii) **Titration with 0.1N $KMnO_4$:** The burette is washed with distilled water. It is rinsed with 0.1N $KMnO_4$ and filled with the same. Initial reading of the burette is noted. Now 25 c.c. of acid mixture is pipetted out and taken into the conical flask. It is heated on the wire gauze to 60–70°C. The 0.1N $KMnO_4$ solution is run from the burette into the conical flask. The conical flask is shaken vigorously and constantly. At the end point, the solution of the conical flask becomes pink. The final reading of the burette is noted. The titration is repeated till two concordant readings are obtained.

Observation and Calculation:

Table 17

Sl. No.	Volume of 0.1N NaOH (c.c.)	Volume of acid mix. taken in burette (c.c.)			Concordant reading (c.c.) (say)	Indicator	End point
		Initial reading (say)	Final reading (say)	Diff. (say)			
1.	25	3.9	28.6	24.7			
2.	25	0.5	25.1	24.6	24.6	Phenolphthalein	Colourless
3.	25	25.1	49.7	24.6			

24.6 c.c. acid mix ≡ 25 c.c. 0.1N NaOH

or, $V_1N_1 = V_2N_2$

or, $24.6 \times N_1 = 25 \times 0.1$

or, $N_1 = \dfrac{25 \times 0.1}{24.6}$

or, $N_1 = 0.102$

∴ Normality of the acid mix. = 0.102 ... (1)

Table 18

Sl. No.	Volume of acid mix. (c.c.)	Volume of 0.1N $KMnO_4$ taken in burette (c.c.)			Concordant reading (c.c.) (say)	Indicator	End point
		Initial reading (say)	Final reading (say)	Diff. (say)			
1.	25	0.9	12.0	11.1			
2.	25	12.0	23.0	11.0	11.0	Self indicator	Pink
3.	25	23.0	34.0	11.0			

25 c.c. oxalic acid ≡ 11 c.c. 0.1N $KMnO_4$

or, $V_1N_1 = V_2N_2$

or, $25 \times N_1 = 11 \times 0.1$

or, $N_1 = \dfrac{11 \times 0.1}{25}$

or, $N_1 = 0.044$

∴ Normality of oxalic acid = 0.044 ... (2)

∴ Gram/litre of hydrated oxalic acid

= Normality × Eq. wt.

= 0.044 × 63 | Mol. wt. of $H_2C_2O_4 \cdot 2H_2O$ = 126

= 2.772 | Eq. wt. of $H_2C_2O_4 \cdot 2H_2O = \frac{126}{2} = 63$

Normality of sulphuric acid

= 0.102 – 0.044 | From equations (1) and (2)

= 0.058

∴ Gram/litre of sulphuric acid

= Normality × Eq. wt. | Mol. wt. of H_2SO_4 = 98

= 0.058 × 49 | Eq. wt. of H_2SO_4 = 49

= 2.842

14.6. To estimate the strength of oxalic acid and sodium oxalate in a mixture of the two using 0.05N NaOH and 0.1N $KMnO_4$

Theory: The mixture solution is first titrated with 0.05N NaOH solution. It will estimate only the amount of oxalic acid.

$$\begin{matrix} COOH \\ | \\ COOH \end{matrix} + 2NaOH \longrightarrow \begin{matrix} COONa \\ | \\ COONa \end{matrix} + 2H_2O$$

In second titration, the mixture solution is titrated with 0.1N $KMnO_4$. It will estimate both component of the mixture.

$$\left[MnO_4^- + 8H^+ + 5\cancel{e} \rightarrow Mn^{2+} + 4H_2O\right] \times 2$$

$$\left[C_2O_4^{2-} \rightarrow 2CO_2 + 2\cancel{e}\right] \times 5$$

$$2MnO_4^- + \underset{\text{(Oxalate/oxalic acid)}}{5C_2O_4^{2-}} + 16H^+ \rightarrow 2Mn^{2+} + 8H_2O + 10CO_2$$

Procedure: It involves following steps:

(i) Titration with 0.05N NaOH: The burette is rinsed with the mixture solution and filled with the same. Initial reading of the burette is noted. Now 25 c.c. of 0.05N NaOH solution is pipetted out and taken into a conical flask. 2-3 drops of phenolphthalein are added to it. The mixture solution is run from the burette into the conical flask. The conical flask is stirred constantly. At the end point, the solution becomes colourless. Final reading of the burette is noted.

(ii) **Titration with 0.1N $KMnO_4$:** The burette is washed with distilled water. It is rinsed with 0.1N $KMnO_4$ and filled with the same. Initial reading of the burette is noted. Now the mixture solution is pipetted out and taken into the conical flask. One test tube of dilute H_2SO_4 is added to it. It is heated on the wire gauze to 60–70°C. Then 0.1N $KMnO_4$ solution is run from the burette into the conical flask. At the end point, the solution becomes pink. The final reading of the burette is noted.

Observation and Calculation:

Table 19

Sl. No.	Vol. of 0.05N NaOH (c.c.)	Volume of mixture solution taken in burette (c.c.)			Concordant reading (c.c.) (say)	Indicator	End point
		Initial reading (say)	Final reading (say)	Diff. (say)			
1.	25	0.0	24.3	24.3			
2.	25	24.3	48.5	24.2	24.2	Phenolphthalein	Colourless
3.	25	0.2	24.4	24.2			

$$24.2 \text{ c.c. oxalic acid} \equiv 25 \text{ c.c. } 0.05\text{N NaOH}$$

or, $V_1N_1 = V_2N_2$

or, $24.2 \times N_1 = 25 \times 0.05$

or, $N_1 = \dfrac{25 \times 0.05}{24.2}$

or, $N_1 = 0.051$

$\therefore$ Normality of oxalic acid = 0.051 ... (1)

∴ Gram/litre of hydrated oxalic acid

= Normality × Eq. wt.
= 0.051 × 63
= 3.21

Mol. wt. of hydrated oxalic acid = 126

$$\text{Eq. wt.} = \frac{126}{2} = 63$$

Table 20

Sl. No.	Vol. of mixture solution (c.c.)	Vol. of 0.1N $KMnO_4$ taken in burette (c.c.)			Concordant reading (c.c.) (say)	Indicator	End point
		Initial reading (say)	Final reading (say)	Diff. (say)			
1.	25	0.0	25.0	25.0			
2.	25	2.2	27.0	24.8	24.8	Self indicator	Pink
3.	25	3.5	28.3	24.8			

25 c.c. oxalic acid and oxalate ≡ 24.8 c.c. 0.1N $KMnO_4$

or, $V_1N_1 = V_2N_2$

or, $25 \times N_1 = 24.8 \times 0.1$

or, $N_1 = \dfrac{24.8 \times 0.1}{25}$

or, $N_1 = 0.099$

∴ Normality of oxalic acid and oxalate = 0.099 ... (2)

∴ Normality of oxalate = 0.099 – 0.051
= 0.048

∴ Gram/litre of sodium oxalate

= Normality × Eq. wt.
= 0.048 × 67
= 3.22

Mol. wt. of sod. oxalate = 134

$$\text{Eq. wt.} = \frac{134}{2} = 67$$

14.7. To determine the percentage purity of given iron wire using 0.1N $KMnO_4$

Theory: A small piece of the given iron wire is weighed accurately and dissolved in dilute H_2SO_4. Thus we get a solution of ferrous sulphate.

$Fe \rightarrow Fe^{2+} + 2e$

The volume of the ferrous sulphate solution thus obtained is made 250 c.c. Now 25 c.c. of this solution is titrated with 0.1N $KMnO_4$.

$$MnO_4^- + 8H^+ + \cancel{5e} \rightarrow Mn^{2+} + 4H_2O$$

$$\left[Fe^{2+} \rightarrow Fe^{3+} + \cancel{e}\right] \times 5$$

$$\overline{MnO_4^- + 5Fe^{2+} + 8H^+ \rightarrow Mn^{2+} + 5Fe^{3+} + 4H_2O}$$

Procedure: It involves following steps:

(i) To dissolve iron wire: At first the given iron wire is weighed accurately. It is taken into a R.B. flask and about 75 c.c. dilute sulphuric acid and 1 c.c. concentrated sulphuric acid are added to it. The R.B. flask fitted with Bunsen valve is heated slowly until the wire dissolves. Now the solution is completely transferred into a 250 c.c. measuring flask and the volume of the solution is made up to the mark.

(ii) Titration: The burette is rinsed with 0.1N $KMnO_4$ and filled with the same. Initial reading is noted. Now 25 c.c. of ferrous sulphate solution (which has been prepared above) is pipetted out and taken into a conical flask. A test tube of dilute H_2SO_4 is added to it and heated on a wire gauze to 60–70°C. The 0.1N $KMnO_4$ solution is run from the burette into the conical flask. The conical flask is constantly shaken during the addition of 0.1N $KMnO_4$ solution. At the end point, the solution becomes pink. The final reading of the burette is noted.

Observation and Calculation:

Wt. of iron wire = 0.887 g (say)

This wire is dissolved in dil. H_2SO_4 and the volume is made 250 c.c.

Table 21

Sl. No.	Vol. of Fe^{2+} solution (c.c.)	Vol. of 0.1N $KMnO_4$ taken in burette (c.c.)			Concordant reading (c.c.) (say)	Indicator	End point
		Initial reading (say)	Final reading (say)	Diff. (say)			
1.	25	0.7	16.2	15.5			
2.	25	1.2	16.6	15.4	15.4	Self indicator	Pink
3.	25	4.6	20.0	15.4			

25 c.c. Fe^{2+} solution ≡ 15.4 c.c. 0.1N $KMnO_4$

or, $V_1N_1 = V_2N_2$

or, $25 \times N_1 = 15.4 \times 0.1$

or, $N_1 = \dfrac{15.4 \times 0.1}{25}$

or, $N_1 = 0.061$

∴ Normality of Fe^{2+} ion = 0.061

∴ Gram/litre of Fe^{2+} ion

= Normality × Eq. wt.

= 0.061 × 55.85

= 3.41

$Fe^{2+} \rightarrow Fe^{3+} + e$

$\therefore$ Eq. wt. $= \dfrac{55.85}{1} = 55.85$

∵ 1000 c.c. Fe^{2+} solution contains 3.41 g iron

∴ 250 c.c. Fe^{2+} solution contains $\dfrac{3.41 \times 250}{1000} = 0.853$ g

∵ 0.887 g iron wire contains 0.853 g iron

∴ 100 g iron wire contains $\dfrac{0.853 \times 100}{0.887} = 96.17$ g iron

Hence, the given iron wire is 96.17% pure.

14.8. To estimate ferric ion in ferric ammonium sulphate solution using 0.1N $KMnO_4$

Theory: First of all ferric ion is reduced into ferrous ion by stannous chloride.

$$\left[Fe^{3+} + \cancel{e} \rightarrow Fe^{2+}\right] \times 2$$

$$Sn^{2+} \rightarrow Sn^{4+} + \cancel{2e}$$

$$2Fe^{3+} + Sn^{2+} \rightarrow 2Fe^{2+} + Sn^{4+}$$

Excess of stannous chloride must be removed completely as it will also be oxidised by $KMnO_4$ along with ferrous ion. This is done by means of mercuric chloride (corrosive sublimate).

$$Sn^{2+} \rightarrow Sn^{4+} + 2\cancel{e}$$

$$2Hg^{2+} + 2\cancel{e} \rightarrow Hg_2^{2+}$$

$$Sn^{2+} + 2Hg^{2+} \rightarrow Sn^{4+} + Hg_2^{2+}$$

At last the ferrous ion is titrated with 0.1N $KMnO_4$ in presence of Zimmermann-Reinhardt solution which prevents oxidation of chloride ions present in the solution.

Procedure: It involves following steps:

(i) **Conversion of ferric into ferrous ions:** First of all 25 c.c. of Fe^{3+} solution is pipetted out and taken into a conical flask. 10 c.c. of concentrated hydrochloric acid is added to it and heated to boiling. Now stannous chloride solution is added to it dropwise with the help of a burette until the yellow colour of Fe^{3+} solution changes to faint green colour of Fe^{2+} solution.

The conical flask is cooled rapidly under tap water (to protect the solution from oxidation by air a cork must be fitted in the mouth of the flask). Then 100 c.c. of distilled water and 10 c.c. of mercuric chloride (corrosive sublimate) solution are added to it. On adding mercuric chloride, a white silky precipitate of mercurous chloride (calomel) appears. This slight amount of mercurous ions has not appreciable effect on oxidising agent ($KMnO_4$). Again 100 c.c. of distilled water is added. Then 20 c.c. of Zimmermann-Reinhardt solution (preventing solution) is added to it.

(ii) **Titration:** Now the whole of the solution of the conical flask is titrated with 0.1N $KMnO_4$ (taken in the burette). At the end point, the solution becomes pink. The whole process is repeated till two concordant readings are obtained.

Observation and Calculation:

Table 22

Sl. No.	Vol. of the supplied soln. (c.c.)	Vol. of 0.1N $KMnO_4$ taken in burette (c.c.)			Concordant reading (c.c.) (say)	Indicator	End point
		Initial reading (say)	Final reading (say)	Diff. (say)			
1.	25	0.55	24.95	24.40			
2.	25	2.20	26.40	24.20	24.2	Self indicator	Pink
3.	25	2.00	26.20	24.20			

$$25 \text{ c.c. } Fe^{2+} \text{ solution} \equiv 24.2 \text{ c.c. } 0.1N\ KMnO_4$$

or, $25 \text{ c.c. } Fe^{3+} \text{ solution} \equiv 24.2 \text{ c.c. } 0.1N\ KMnO_4$

or, $V_1N_1 = V_2N_2$

or, $25 \times N_1 = 24.2 \times 0.1$

or, $N_1 = \dfrac{24.2 \times 0.1}{25}$

or, $N_1 = 0.097$

$\therefore$ Gram/litre of Fe^{3+} ion

$= \text{Normality} \times \text{Eq. wt.}$

$= 0.097 \times 55.85$

$= 5.42$

$Fe^{2+} \rightarrow Fe^{3+} + e$

$\therefore \text{Eq. wt.} = \dfrac{55.85}{1}$

$= 55.85$

Things to Remember

By using Zimmermann-Reinhardt solution (Preventing solution), Fe^{2+} ion can be estimated by $KMnO_4$ in presence of Cl^- ions.

Zimmermann-Reinhardt solution is prepared by dissolving 50 g crystalline manganese sulphate ($MnSO_4 \cdot 4H_2O$) in 200 c.c. water and adding a cooled mixture of 100 c.c. conc. H_2SO_4 and 300 c.c. water followed by 100 c.c. syrupy phosphoric acid. Main functions of this solution may be described as under:

(i) Permanganate ion is reduced as under:

$$MnO_4^- + 8H^+ + 5e \rightarrow Mn^{2+} + 4H_2O$$

The reduction potential (E) of this system is given by the Nernst equation:

$$E = E^\circ + \frac{0.0591}{5} \log \frac{[MnO_4^-][H^+]^8}{[Mn^{2+}]} \text{ at } 25°C$$

$[H_2O]$ = a constant

Concentration of electron = 1

$$MnSO_4 \rightleftharpoons Mn^{2+} + SO_4^{2-}$$

Since Zimmermann-Reinhardt solution furnishes Mn^{2+} ions, hence $[Mn^{2+}]$ increases and thereby the reduction potential of the couple MnO_4^-/Mn^{2+} decreases. Hence the oxidising power of MnO_4^- decreases. Thus Zimmermann-Reinhardt solution decreases the oxidising power of MnO_4^- ion. In this way MnO_4^- cannot oxidise Cl^- into Cl_2 in presence of the preventing solution.

(ii) Phosphoric acid reacts with Fe^{3+} ion to form a complex ion.

$$Fe^{3+} + H_3PO_4 \rightleftharpoons [Fe(HPO_4)]^+ + 2H^+$$

Thus phosphoric acid reduces the concentration of Fe^{3+} ion.

Ferric ion is reduced to ferrous ion as under:

$$Fe^{3+} + e \rightarrow Fe^{2+}$$

Reduction potential of the couple Fe^{3+}/Fe^{2+} is given by Nernst equation:

$$E = E^\circ + \frac{0.0591}{1} \log \frac{[Fe^{3+}]}{[Fe^{2+}]} \text{ at } 25°\,C$$

Thus due to decrease in concentration of Fe^{3+} ion, reduction potential of the couple Fe^{3+}/Fe^{2+} is lowered. Hence reducing power of Fe^{2+} ion is increased. Under this condition, permanganate ion can oxidise ferrous ion easily.

(iii) Fe^{3+} ion is yellow in colour. It reacts with phosphoric acid to form a colourless complex ion. Thus at the end point, the purple/pink colouration of $KMnO_4$ is clearly visible.

$$\underset{(Yellow)}{Fe^{3+}} + H_3PO_4 \rightleftharpoons \underset{(Colourless)}{[Fe(HPO_4)]^+} + 2H^+$$

14.9. To estimate Fe^{2+} and Fe^{3+} ions in the given solution using 0.1N $KMnO_4$

Theory: Ferrous ion of the mixture solution is directly estimated by titrating it with 0.1N $KMnO_4$.

$$MnO_4^- + 8H^+ + 5\cancel{e} \rightarrow Mn^{2+} + 4H_2O$$

$$[Fe^{2+} \rightarrow Fe^{3+} + \cancel{e}] \times 5$$

$$\overline{MnO_4^- + 5Fe^{2+} + 8H^+ \rightarrow Mn^{2+} + 5Fe^{3+} + 4H_2O}$$

For estimation of ferric ion present in the mixture solution, ferric ion is first reduced to ferrous ion by $SnCl_2$.

$$[Fe^{3+} + \cancel{e} \rightarrow Fe^{2+}] \times 2$$

$$Sn^{2+} \rightarrow Sn^{4+} + 2\cancel{e}$$

$$\overline{2Fe^{3+} + Sn^{2+} \rightarrow 2Fe^{2+} + Sn^{4+}}$$

Excess of $SnCl_2$ must be removed completely as it is oxidised by $KMnO_4$ solution. This is done by $HgCl_2$ which reacts with $SnCl_2$ as shown below:

$$Sn^{2+} \rightarrow Sn^{4+} + 2\cancel{e}$$

$$2Hg^{2+} + \cancel{2e} \rightarrow Hg_2^{2+}$$

$$\overline{Sn^{2+} + 2Hg^{2+} \rightarrow Sn^{4+} + Hg_2^{2+}}$$

Now the total Fe^{2+} ion (Original + reduced) is estimated by titrating with 0.1N $KMnO_4$.

$$MnO_4^- + 8H^+ + \cancel{5e} \rightarrow Mn^{2+} + 4H_2O$$

$$[Fe^{2+} \rightarrow Fe^{3+} + \cancel{e}] \times 5$$

$$\overline{MnO_4^- + 5Fe^{2+} + 8H^+ \rightarrow Mn^{2+} + 5Fe^{3+} + 4H_2O}$$

Procedure: It involves following steps:

(i) **First titration:** The burette is rinsed with 0.1N $KMnO_4$ and filled with the same. Initial reading of the burette is noted.

Now 25 c.c. of mixture solution is pipetted out and taken into the conical flask. One test tube of dilute sulphuric acid is added to it. 0.1N $KMnO_4$ solution is run from the burette into the conical flask. The flask is shaken constantly during the addition of the $KMnO_4$. At the end point, the colour of the solution becomes pink. The final reading of the burette is noted. The process is repeated till two concordant readings are obtained. From this titration, ferrous ion can be estimated.

(ii) **Second titration:** 25 c.c. of the mixture solution is pipetted out and taken into a conical flask. 10 c.c. concentrated hydrochloric acid is added to it and heated to boil. Now stannous chloride solution is added to it dropwise from a burette till the yellow colour of the ferric ion turns to faint green (due to Fe^{2+} ions). The flask is then cooled under tap water. Now 100 c.c. of distilled water is added to it followed by 10 c.c. of $HgCl_2$ solution.

On addition of $HgCl_2$ solution, white silky precipitate of Hg_2Cl_2 appears. This slight amount of mercurous ion has no appreciable effect on oxidising agent like $KMnO_4$. Again 100 c.c. of distilled water is added followed by 20 c.c. of Zimmermann-Reinhardt solution. Now the whole solution is titrated with 0.1N $KMnO_4$. In this titration, potassium permanganate oxidises the whole of the ferrous ion (original ferrous ion and reduced ferrous ion).

Observation and Calculation:

Table 23

Sl. No.	Vol. of mix soln. (Fe^{2+} and Fe^{3+}) (c.c.)	Vol. of 0.1N $KMnO_4$ taken in burette (c.c.)			Concordant reading (c.c.) (say)	Indicator	End point
		Initial reading (say)	Final reading (say)	Diff. (say)			
1.	25	6.00	19.80	13.80			
2.	25	5.40	18.75	13.35	13.35	Self indicator	Pink
3.	25	2.70	16.05	13.35			

25 c.c. Fe^{2+} ion (original) $\equiv$ 13.35 c.c. 0.1N $KMnO_4$

or, $V_1N_1 = V_2N_2$

or, $25 \times N_1 = 13.35 \times 0.1$

or, $N_1 = \dfrac{13.35 \times 0.1}{25}$

or, $N_1 = 0.053$

$\therefore$ Normality of Fe^{2+} ion = 0.053 ... (1)

$\therefore$ Gram/litre of Fe^{2+} ion

= Normality × Eq. wt.

= 0.053 × 55.85

= 2.96

$Fe^{2+} \rightarrow Fe^{3+} + e$

$\therefore$ Eq. wt. $= \dfrac{55.85}{1} = 55.85$

Table 24

Sl. No.	Vol. of mix soln. (original Fe^{2+} + reduced Fe^{2+}) (c.c.)	Vol. of 0.1N $KMnO_4$ taken in burette (c.c.)			Concordant reading (c.c.) (say)	Indicator	End point
		Initial reading (say)	Final reading (say)	Diff. (say)			
1.	25	0.30	28.20	27.90			
2.	25	1.50	29.30	27.80	27.80	Self indicator	Pink
3.	25	2.40	30.20	27.80			

25 c.c. Fe^{2+} ion (original + reduced)

$\equiv$ 27.80 c.c. 0.1N $KMnO_4$

or, $V_1N_1 = V_2N_2$

or, $25 \times N_1 = 27.80 \times 0.1$

or, $N_1 = \dfrac{27.80 \times 0.1}{25}$

or, $N_1 = 0.111$

$\therefore$ Normality of Fe^{2+} ion (original + reduced)

= 0.111 ... (2)

$\therefore$ Normality of Fe^{3+} ion

= Normality of Fe^{2+} ion (original + reduced)

– Normality of Fe^{2+} ion (original)

$= 0.111 - 0.053$

$= 0.058$

$\therefore$ Gram/litre of Fe^{3+} ion

$= \text{Normality} \times \text{Eq. wt.}$

$= 0.058 \times 55.85$

$= 3.24$

$Fe^{2+} \rightarrow Fe^{3+} + e$

$\therefore \text{Eq. wt.} = \frac{55.85}{1} = 55.85$

14.10. To estimate calcium in calcium carbonate with the help of standard $KMnO_4$

Theory: Weighed amount of calcium carbonate is dissolved in dilute hydrochloric acid.

$$CaCO_3 + 2H^+ \rightarrow Ca^{2+} + H_2O + CO_2 \uparrow \quad \text{... (1)}$$

The calcium ion is precipitated out as calcium oxalate by adding ammonium oxalate.

$$Ca^{2+} + C_2O_4^{2-} \rightarrow CaC_2O_4 \downarrow \quad \text{... (2)}$$

From equation (2), we can write,

$$Ca^{2+} \equiv C_2O_4^{2-} \quad \text{... (3)}$$

The precipitate of calcium oxalate is then dissolved in dilute sulphuric acid.

$$CaC_2O_4 + 2H^+ \rightarrow Ca^{2+} + H_2C_2O_4 \quad \text{... (4)}$$

Now the liberated oxalic acid is titrated with 0.1N $KMnO_4$.

$$\left[MnO_4^- + 8H^+ + \cancel{5e} \rightarrow Mn^{2+} + 4H_2O\right] \times 2 \quad \text{... (5)}$$

$$\left[C_2O_4^{2-} \rightarrow 2CO_2 + 2\cancel{e}\right] \times 5 \quad \text{... (6)}$$

$$2MnO_4^- + 5C_2O_4^{2-} + 16H^+ \rightarrow 2Mn^{2+} + 8H_2O + 10CO_2 \quad \text{... (7)}$$

From relation (3) and (6), we can write,

$$Ca^{2+} \equiv C_2O_4^{2-} \equiv 2e \quad \text{... (8)}$$

$\therefore$ Eq. wt. of $Ca^{2+} = \frac{40}{2}$

$= 20$

Procedure: It involves following steps:

(i) **Preparation of solution of $CaCO_3$:** First of all 0.172 g of $CaCO_3$ is weighed and taken into a beaker. 20 c.c. distilled water and 10 c.c. dilute hydrochloric acid (1:1) are added to it. The beaker is heated until calcium carbonate dissolves. The solution is diluted to 200 c.c. and few drops of methyl red are added. It is heated to boiling and then a clear solution of 6% ammonium oxalate is added till complete precipitation of calcium oxalate. The solution is heated to 70–80°C. Then dilute ammonia (1:1) solution is added dropwise with stirring until the solution becomes neutral or slightly alkaline. It is indicated by the colour change of the solution from red to yellow. The solution is allowed to stand and the clear supernatant liquid decanted through Whatman filter paper Number 40. The precipitate is washed with the help of distilled water to remove ammonium oxalate and chloride completely. The presence of oxalate and chloride is tested by adding $AgNO_3$ solution in the filtrate (washing). If oxalate or chloride is present, white precipitate appears on addition of $AgNO_3$ solution in the washing. In this way oxalate and chloride are completely removed from the precipitate of calcium oxalate. This precipitate is now quantitatively dissolved in dilute H_2SO_4 (1:8). This solution is taken into a 250 c.c. measuring flask and the volume is made up to the mark.

(ii) **Titration:** The burette is rinsed with 0.1N $KMnO_4$ and filled by it. Initial reading of the burette is noted. Now 25 c.c. of the above prepared calcium oxalate solution is pipetted out and taken into a flask. It is heated on a wire gauze to 60–70°C. Then 0.1N $KMnO_4$ solution is run from burette into the conical flask. At the end point, the solution becomes pink. Final reading of the burette is noted.

Observation and Calculation:

Wt. of given $CaCO_3$ = 0.172 g (say)

Table 25

Sl. No.	Vol. of liberated oxalic acid (c.c.)	Vol. of 0.1N $KMnO_4$ taken in burette (c.c.)			Concordant reading (c.c.) (say)	Indicator	End point
		Initial reading (say)	Final reading (say)	Diff. (say)			
1.	25	1.95	5.60	3.65			
2.	25	5.65	9.20	3.55	3.55	Self indicator	Pink
3.	25	9.20	12.75	3.55			

25 c.c. oxalic acid ≡ 3.55 c.c. 0.1N $KMnO_4$

or, 25 c.c. Ca^{2+} ion ≡ 3.55 c.c. 0.1N $KMnO_4$

| From relation (3)

or, $V_1N_1 = V_2N_2$

or, $25 \times N_1 = 3.55 \times 0.1$

or, $N_1 = \dfrac{3.55 \times 0.1}{25}$

or, $N_1 = 0.014$

∴ Normality of Ca^{2+} ion = 0.014

∴ Gram/litre of Ca^{2+} ion

= Normality × Eq. wt.

= 0.014 × 20

= 0.28

∵ 1000 c.c. solution of Ca^{2+} ion contains 0.28 g of Ca^{2+} ion

∴ 250 c.c. solution of Ca^{2+} ion contains $\dfrac{0.28 \times 250}{1000}$

= 0.07 g of Ca^{2+} ion

∵ 0.172 g of given $CaCO_3$ contains 0.07 g of Ca^{2+} ion

∴ 100 g of given $CaCO_3$ contains $\dfrac{0.07 \times 100}{0.172}$

= 40.7 g of Ca^{2+} ion

Therefore, amount of calcium in given $CaCO_3$ sample = 40.7%.

Things to Remember

Calcium/calcium carbonate in chalk may be estimated using similar method, but one thing should be noted here:

After dissolving the chalk in dilute hydrochloric acid, it must be filtered and the experiment is proceeded from filtrate.

14.11. To determine the number of molecules of water of crystallisation in a molecule of the given sample of hydrated oxalic acid using 0.1N $KMnO_4$

Theory: A known weight of hydrated oxalic acid is dissolved in water. Then a known volume of this is titrated with $KMnO_4$ in presence of dilute sulphuric acid. From this titration, the gram per litre of anhydrous oxalic acid can be calculated. Then by applying the following relation, the number of molecules of water of crystallisation may be determined:

$$\frac{\text{Gram/litre of hydrated salt}}{\text{Gram/litre of anhydrous salt}} = \frac{\text{Mol. wt. of hydrated salt}}{\text{Mol. wt. of anhydrous salt}}$$

The chemical reaction occurring in this titration can be written as,

$$\left[MnO_4^- + 8H^+ + 5\cancel{e} \rightarrow Mn^{2+} + 4H_2O\right] \times 2$$

$$\left[C_2O_4^{2-} \rightarrow 2CO_2 + 2\cancel{e}\right] \times 5$$

$$2MnO_4^- + 5C_2O_4^{2-} + 16H^+ \rightarrow 2Mn^{2+} + 8H_2O + 10CO_2$$

Procedure: It involves following steps:

(i) Preparation of solution of oxalic acid: First of all 1.5 g of hydrated oxalic acid (given sample) is weighed and taken into a 250 c.c. measuring flask. To this 100 c.c. of distilled water is added and shaken to dissolve the acid. The volume of the solution is made up to the mark on the measuring flask with distilled water.

(ii) Titration: The burette is rinsed with 0.1N $KMnO_4$ and filled by it. Initial reading of the burette is noted. Now 25 c.c. of the solution (prepared as above) is pipetted out and taken into a conical flask. The flask is heated on a wire gauze to 60–70°C. Then 0.1N $KMnO_4$ is run from burette into the flask. At the end point, the solution becomes pink. The final reading of the burette is noted.

Observation and Calculation:

Table 26

Sl. No.	Vol. of the acid soln. (c.c.)	Vol. of 0.1N $KMnO_4$ taken in burette (c.c.)			Concordant reading (c.c.) (say)	Indicator	End point
		Initial reading (say)	Final reading (say)	Diff. (say)			
1.	25	0.0	24.6	24.6			
2.	25	24.6	49.1	24.5	24.5	Self indicator	Pink
3.	25	2.0	26.5	24.5			

25 c.c. acid solution ≡ 24.5 c.c. 0.1N $KMnO_4$

or, $V_1N_1 = V_2N_2$

or, $25 \times N_1 = 24.5 \times 0.1$

or, $N_1 = \dfrac{24.5 \times 0.1}{25}$

or, $N_1 = 0.098$

∴ Normality of anhydrous oxalic acid = 0.098

∴ Gram/litre of anhydrous oxalic acid

$= \text{Normality} \times \text{Eq. wt.}$

$= 0.098 \times 45$

$= 4.41$

Eq. wt. of $H_2C_2O_4 = \dfrac{90}{2} = 45$

Now from theory:

$$\frac{\text{Gram/litre of hydrated oxalic acid}}{\text{Gram/litre of anhydrous oxalic acid}} = \frac{\text{Mol. wt. of hydrated oxalic acid}}{\text{Mol. wt. of anhydrous oxalic acid}}$$

or, $\dfrac{1.5 \times 4}{4.41} = \dfrac{90 + 18x}{90}$

or, $\dfrac{6}{4.41} = \dfrac{90 + 18x}{90}$

or, $6 \times 90 = 4.41 \times 90 + 4.41 \times 18x$

or, $540 = 396.9 + 79.38 \times x$

or, $79.38 \times x = 540 - 396.9$

or, $79.38 \times x = 143.1$

or, $x = \frac{143.1}{79.38}$

or, $x = 1.80$

or, $x = 2$ (nearest whole number)

Thus the number of molecules of water of crystallisation in hydrated oxalic acid (given sample) is **2.**

14.12. To prepare 0.1N $K_2Cr_2O_7$ and to determine the strength of given Mohr's salt solution

Theory: $Cr_2O_7^{2-} + 14H^+ + 6e \rightarrow 2Cr^{3+} + 7H_2O$

$$\therefore \text{ Eq. wt. of } K_2Cr_2O_7 = \frac{\text{Mol. wt. of } K_2Cr_2O_7}{\text{Number of electrons gained}}$$

$$= \frac{39 \times 2 + 52 \times 2 + 16 \times 7}{6}$$

$$= \frac{294}{6}$$

$$= 49$$

$\because$ For 1000 c.c. N solution, 49 g of $K_2Cr_2O_7$ is needed.

$\therefore$ For 1000 c.c. 0.1N solution, 49×0.1 g of $K_2Cr_2O_7$ is needed.

$\therefore$ For 250 c.c. 0.1N solution, $\frac{49 \times 0.1 \times 250}{1000}$ g of $K_2Cr_2O_7$ is needed.

= 1.225 g of $K_2Cr_2O_7$ is needed.

Thus 1.225 g potassium dichromate is dissolved in water and the volume is made 250 c.c. to get 0.1N $K_2Cr_2O_7$ solution.

Then the given Mohr's salt solution is titrated against 0.1N $K_2Cr_2O_7$.

$$Cr_2O_7^{2-} + 14H^+ + 6\not{e} \rightarrow 2Cr^{3+} + 7H_2O$$

$$\left[Fe^{2+} \rightarrow Fe^{3+} + \not{e}\right] \times 6$$

$$\underset{\text{(Light yellow)}}{Cr_2O_7^{2-}} + 6Fe^{2+} + 14H^+ \rightarrow \underset{\text{(Green)}}{2Cr^{3+}} + 6Fe^{3+} + 7H_2O$$

Procedure: It involves following steps:

(i) **Preparation of 0.1N $K_2Cr_2O_7$:** First of all 1.225 g of $K_2Cr_2O_7$ is weighed and taken into a 250 c.c. measuring flask. 50 c.c. of hot water is added to it. The flask is shaken to dissolve the potassium dichromate. Now the volume of the solution is made 250 c.c. at room temperature by adding distilled water. Now 0.1N $K_2Cr_2O_7$ is ready.

(ii) **Titration:** The burette is rinsed with 0.1N $K_2Cr_2O_7$ and filled by it. Initial reading of the burette is noted. 25 c.c. of Mohr's salt solution is pipetted out and taken into a conical flask. Two test tubes of bench sulphuric acid, 5 c.c. of syrupy phosphoric acid, and 8 drops of sodium diphenylamine sulphonate are added. Now 0.1N potassium dichromate is run from the burette into the conical flask. At the end point, the green colour of the solution due to Fe^{2+} ions changes into bluish-violet. Final reading of the burette is noted. The process is repeated till two concordant readings are obtained.

Observation and Calculation:

1.225 g of $K_2Cr_2O_7$ is dissolved in water and the volume is made 250 c.c. to get 0.1N $K_2Cr_2O_7$.

Table 27

Sl. No.	Vol. of Mohr's salt soln. (c.c.)	Vol. of 0.1N $K_2Cr_2O_7$ taken in burette (c.c.)			Concordant reading (c.c.) (say)	Indicator	End point
		Initial reading (say)	Final reading (say)	Diff. (say)			
1.	25	1.10	26.40	25.30			
2.	25	0.95	26.15	25.20	25.20	Sodium diphenyl-amine sulphonate	Bluish-violet
3.	25	1.60	26.80	25.20			

25 c.c. Mohr's salt solution ≡ 25.2 c.c. 0.1N $K_2Cr_2O_7$

or, $V_1N_1 = V_2N_2$

or, $25 \times N_1 = 25.2 \times 0.1$

or, $N_1 = \frac{25.2 \times 0.1}{25}$

or, $N_1 = 0.1008$

∴ Normality of Mohr's salt = 0.1008

∴ Gram/litre of Mohr's salt

= Normality × Eq. wt.

= 0.1008 × 392

= 39.5

$Fe^{2+} \rightarrow Fe^{3+} + e$

∴ Eq. wt. of Mohr's salt $= \frac{392}{1} = 392$

Things to Remember

(i) The titration which involves the use of potassium dichromate solution is known as dichromatomerty.

(ii) Potassium dichromate is not so strong oxidant but has many useful applications and certain advantages over the other oxidants like potassium permanganate, ceric sulphate, etc. Some of these advantages are:

- Potassium dichromate can be obtained in a very high purity and, therefore, serves as an excellent primary standard.
- Its solution is stable indefinitely if it is properly protected from evaporation.
- Its solution is stable towards light and is little affected by dust and organic matter.
- It can be used as an oxidising agent in presence of HCl or chloride. Its solution does not oxidise chlorides at room temperature even in 1–2N acid. There is, however, some reduction of dichromate by chloride at temperature close to the boiling point.

(iii) There are certain disadvantages of potassium dichromate over potassium permanganate.

- Since dichromate is reduced to a highly coloured chromic ion, it is impossible to observe the end point of these titrations visually without an added indicator. In fact, the colour of the reduced ion even with an added indicator makes the detection of the end point difficult for some people.
- The oxidant has limited applications and its titrations are comparatively tedious.

(iv) Potassium dichromate is a weak oxidising agent. To explain this, standard reduction potentials of few systems are given below:

Couple	*Standard reduction potential (E°)*
MnO_4^-/Mn^{2+}	1.52 volt
Cl_2/Cl^-	1.36 volt
$Cr_2O_7^{2-}/Cr^{3+}$	1.33 volt
Fe^{3+}/Fe^{2+}	0.77 volt
I_2/I^-	0.53 volt

Since E° of the couple $Cr_2O_7^{2-}/Cr^{3+}$ is less than that of MnO_4^-/Mn^{2+}, hence $K_2Cr_2O_7$ is less powerful oxidising agent than $KMnO_4$.

Since E° of the couple MnO_4^-/Mn^{2+} is greater than that of Cl_2/Cl^-, hence MnO_4^- ion oxidises Cl^- ion into Cl_2. $Cr_2O_7^{2-}$ ion does not oxidise Cl^- ion into Cl_2 because E° of $Cr_2O_7^{2-}/Cr^{3+}$ is less than that of Cl_2/Cl^-.

MnO_4^- ion and $Cr_2O_7^{2-}$ ion both oxidise Fe^{2+} ion into Fe^{3+} ion. Hence ferrous ion can be estimated with the help of standard $KMnO_4$ solution and standard $K_2Cr_2O_7$ solution too. Since MnO_4^- ion also oxidises Cl^- ion, hence estimation of Fe^{2+} ion in presence of Cl^- ion by $KMnO_4$ becomes difficult. $Cr_2O_7^{2-}$ ion does not oxidise Cl^- ion and as such Fe^{2+} ion can be estimated in presence of Cl^- ion by $K_2Cr_2O_7$.

(v) Hydrogen ion concentration should be sufficiently high, only then potassium dichromate acts as oxidant. The reduction of dichromate ion can be shown as under:

$$Cr_2O_7^{2-} + 14H^+ + 6e \rightarrow 2Cr^{3+} + 7H_2O$$

The reduction potential of this reaction can be given by the Nernst equation.

$$E = E^\circ + \frac{0.0591}{6} \cdot \log \frac{[Cr_2O_7^{2-}][H^+]^{14}}{[Cr^{3+}]^2} \text{ at } 25^\circ$$

$[H_2O]$ = a const.

Concentration of electron = 1.

From the above equation, it is clear that reduction potential of $Cr_2O_7^{2-}/Cr^{3+}$ couple is directly proportional to concentration of hydrogen. Greater the hydrogen ion concentration, higher will be the reduction potential of $Cr_2O_7^{2-}/Cr^{3+}$ couple. Hence most useful applications of dichromate as an oxidant in volumetric determination are those that are carried in 1–2N acid. It is of no value in neutral or alkaline solution.

(vi) In dichromatometry, it is essential to use indicator. The green colour due to the Cr^{3+} ions formed by the reduction of potassium dichromate makes it impossible to ascertain the end point of a dichromate titration by simple visual inspection of the solution and so a redox indicator must be employed which gives a strong and unmistakable colour change.

(vii) In dichromatometry, the end point may be detected either by means of external indicator (potassium ferricyanide) or internal indicator (diphenylamine). But now-a-days only internal indicator is used.

(viii) Internal indicator is used due to its certain advantages over external indicator. These are:

- When potassium ferricyanide is used as an external indicator, there occurs a loss of titrable Fe^{2+} ions because during the titration some solution is taken out of the flask to test it with the indicator. On the other hand, there is no loss when diphenylamine is used as an indicator.

- Titrations involving potassium ferricyanide as an indicator are tedious and very little use is made of this indicator in present day applications of dichromate. On the other hand, titrations involving diphenylamine are easy to use.

(ix) While titrating ferrous ion with potassium dichromate, addition of syrupy phosphoric acid to the solution containing ferrous ions is essential. This is due to following reasons:

- During titration concentration of ferric ions becomes too high which may cause the oxidation of diphenylamine. This results in premature end point. In order to prevent the premature end point, syrupy phosphoric acid is added which forms a stable complex with ferric ion. This complex is almost undissociated and thus the ferric ions are prevented from oxidising the diphenylamine.

$$Fe^{3+} + H_3PO_4 \rightarrow \underset{\text{(Colourless)}}{\left[Fe(HPO_4)\right]^+} + 2H^+$$

- Syrupy phosphoric acid also increases the reducing power of ferrous ion. The reduction of ferric ion can be shown as below:

$Fe^{3+} + e \rightarrow Fe^{2+}$

The reduction potential for this reaction can be given by Nernst equation:

$$E = E^\circ + \frac{0.0591}{1} \cdot \log \frac{[Fe^{3+}]}{[Fe^{2+}]}$$

Phosphoric acid decreases the concentration of ferric ions. Thus reduction potential of Fe^{3+}/Fe^{2+} couple decreases. This is clear from the Nernst equation. Hence reducing power of ferrous ion increases.

- Phosphoric acid converts yellow coloured Fe^{3+} ions into colourless complex to impart sharp end point.

14.13. To estimate the percentage of ferric ion in the given sample of ferric alum, $Fe_2(SO_4)_3 \cdot (NH_4)_2\ SO_4 \cdot 24H_2O$ using 0.1N $K_2Cr_2O_7$

Theory: Weighed amount of ferric alum is dissolved in water and the volume is made 250 c.c. Ferric ion of the solution (25 c.c.) is reduced to ferrous ion by $SnCl_2$.

$$\left[Fe^{3+} + \cancel{e} \rightarrow Fe^{2+}\right] \times 2$$

$$Sn^{2+} \rightarrow Sn^{4+} + 2\cancel{e}$$

$$\overline{2Fe^{3+} + Sn^{2+} \rightarrow 2Fe^{2+} + Sn^{4+}}$$

Excess of stannous ion is removed by adding mercuric chloride.

$$Sn^{2+} \rightarrow Sn^{4+} + 2\cancel{e}$$

$$2Hg^{2+} + 2\cancel{e} \rightarrow Hg_2^{2+}$$

$$\overline{Sn^{2+} + 2Hg^{2+} \rightarrow Sn^{4+} + Hg_2^{2+}}$$

A little amount of mercurous ions so produced does not affect oxidant. Now the resulting solution is titrated against 0.1N $K_2Cr_2O_7$ using diphenylamine or sodium diphenylamine sulphonate as an indicator.

$$Cr_2O_7^{2-} + 14H^+ + 6\cancel{e} \rightarrow 2Cr^{3+} + 7H_2O$$

$$\left[Fe^{2+} \rightarrow Fe^{3+} + \cancel{e}\right] \times 6$$

$$\overline{Cr_2O_7^{2-} + 6Fe^{2+} + 14H^+ \rightarrow 2Cr^{3+} + 6Fe^{3+} + 7H_2O}$$

Procedure: It involves following steps:

(i) **Preparation of ferric alum solution:** First of all 12 g of ferric alum is exactly weighed and taken into a 250 c.c. measuring flask containing 25 c.c. of dilute sulphuric acid. The flask is shaken to dissolve the solid and the volume is made 250 c.c.

(ii) Titration: The burette is rinsed with 0.1N $K_2Cr_2O_7$ and filled by same. Initial reading is noted. 25 c.c. of ferric alum solution is pipetted out and taken into a conical flask. 3–4 c.c. concentrated hydrochloric acid is added to it and boiled till the solution attains deep yellow colour. Then $SnCl_2$ solution is added dropwise through a pipette till the yellow colour of the solution becomes colourless. The solution is cooled under tap. Now the solution is diluted to 100 c.c. by adding distilled water. To this solution 1–2 c.c. of 5% aq. solution of $HgCl_2$ is added to oxidise the excess $SnCl_2$. If white turbidity appears, it means that the reduction of Fe^{3+} to Fe^{2+} is complete. In case no turbidity appears, the solution is rejected and proceeded again for reduction from the very start. Now it is titrated against 0.1N $K_2Cr_2O_7$ solution using 8 drops of diphenylamine or sodium diphenylamine sulphonate followed by 5 c.c. syrupy phosphoric acid. Final reading of the burette is noted.

Observation and Calculation:

Wt. of ferric alum = 12 g.

This amount is dissolved in water and the volume is made 250 c.c.

Table 28

Sl. No.	Vol. of reduced Fe^{2+} ions (c.c.)	Vol. of 0.1N $K_2Cr_2O_7$ taken in burette (c.c.)			Concordant reading (c.c.) (say)	Indicator	End point
		Initial reading (say)	Final reading (say)	Diff. (say)			
1.	25	0.0	24.1	24.1			
2.	25	3.1	27.1	24.0	24.0	Diphenyl-amine	Bluish-violet
3.	25	5.0	29.0	24.0			

25 c.c. of reduced Fe^{2+} ion ≡ 24 c.c. 0.1N $K_2Cr_2O_7$

or, 25 c.c. of Fe^{3+} ion ≡ 24 c.c. 0.1N $K_2Cr_2O_7$

or, $V_1N_1 = V_2N_2$

or, $25 \times N_1 = 24 \times 0.1$

or, $N_1 = \frac{24 \times 0.1}{25}$

or, $N_1 = 0.096$

$\therefore$ Normality of Fe^{3+} ion = 0.096

$\therefore$ Gram/litre of Fe^{3+} ion

= Normality × Eq. wt.

= 0.096 × 55.85

= 5.36

$Fe^{2+} \rightarrow Fe^{3+} + e$

$\therefore$ Eq. wt. $= \frac{55.85}{1}$

$= 55.85$

$\because$ 1000 c.c. of solution contains 5.36 g of ferric ions

$\therefore$ 250 c.c. of solution contains $\frac{5.36 \times 250}{1000}$ g of ferric ions

= 1.34 g of ferric ions

$\because$ 12 g of given alum contains 1.34 g of ferric ions

$\therefore$ 100 g of given alum contains $\frac{1.34 \times 100}{12}$

= 11.75 g of ferric ions

Therefore, amount of ferric ions in the given alum = 11.75%.

Things to Remember

The stannous chloride solution for reduction is prepared by dissolving 3 g of $SnCl_2 \cdot 2H_2O$ in 40 c.c. of concentrated hydrochloric acid and diluting to 250 c.c.

14.14. To estimate the percentage of Iron in Haematite (Fe_2O_3) using 0.1N $K_2Cr_2O_7$

Theory: Generally, iron ore contains Fe^{2+} and Fe^{3+} ions. Hence ferric ion is first reduced to ferrous ion by $SnCl_2$.

$$\left[Fe^{3+} + \cancel{e} \rightarrow Fe^{2+}\right] \times 2$$

$$Sn^{2+} \rightarrow Sn^{4+} + 2\cancel{e}$$

$$\overline{2Fe^{3+} + Sn^{2+} \rightarrow 2Fe^{2+} + Sn^{4+}}$$

Now slight excess of stannous ion is oxidised by $HgCl_2$.

$$Sn^{2+} \rightarrow Sn^{4+} + 2\not{e}$$
$$2Hg^{2+} + 2\not{e} \rightarrow Hg_2^{2+}$$
$$\overline{Sn^{2+} + 2Hg^{2+} \rightarrow Sn^{4+} + Hg_2^{2+}}$$

Then the total ferrous ion is titrated against 0.1N $K_2Cr_2O_7$.

$$Cr_2O_7^{2-} + 14H^+ + 6\not{e} \rightarrow 2Cr^{3+} + 7H_2O$$
$$\left[Fe^{2+} \rightarrow Fe^{3+} + \not{e}\right] \times 6$$
$$\overline{Cr_2O_7^{2-} + 6Fe^{2+} + 14H^+ \rightarrow 2Cr^{3+} + 6Fe^{3+} + 7H_2O}$$

Procedure: It involves following steps:

(i) **Preparation of solution:** First of all the haematite is dried in an air oven and 2 g of it is weighed. It is taken in a conical flask containing 50 c.c. concentrated hydrochloric acid. The flask is heated gently with occasional shaking to dissolve the ore as much as possible. It is filtered and the residue is washed with dilute HCl. Now the filtrate and washing are taken in a 250 c.c. measuring flask and diluted up to the mark.

(ii) **Titration:** The burette is rinsed with 0.1N $K_2Cr_2O_7$ and filled by it. Initial reading of the burette is noted. Now 25 c.c. of the solution prepared as above is pipetted out and taken into a conical flask. 2 c.c. concentrated hydrochloric acid is added to it and boiled for 2 minutes. Stannous chloride solution is added dropwise through a burette to the hot solution till the yellow colour of the solution gets discharged. Now the solution is cooled under tap water and diluted to about 100 c.c. with distilled water. Then 2 c.c. of saturated mercuric chloride solution is added to it to oxidise excess stannous chloride. A white turbidity is obtained. Now the solution is titrated against 0.1N $K_2Cr_2O_7$ using 8 drops of sodium diphenylamine sulphonate or diphenylamine followed by 5 c.c. syrupy phosphoric acid. At the end point, the solution becomes bluish-violet. Final reading of the burette is noted.

Observation and Calculation:

Wt. of haematite = 2 g.

This amount is dissolved in conc. HCl and the volume is made 250 c.c. with distilled water.

Table 29

Sl. No.	Vol. of solution of haematite (c.c.)	Vol. of 0.1N $K_2Cr_2O_7$ taken in burette (c.c.)			Concordant reading (c.c.) (say)	Indicator	End point
		Initial reading (say)	Final reading (say)	Diff. (say)			
1.	25	0.5	20.6	20.1			
2.	25	20.6	40.6	20.0	20.0	Diphenyl-amine	Bluish-violet
3.	25	5.0	25.0	20.0			

25 c.c. haematite solution ≡ 20 c.c. 0.1N $K_2Cr_2O_7$

or, $V_1N_1 = V_2N_2$

or, $25 \times N_1 = 20 \times 0.1$

or, $N_1 = \dfrac{20 \times 0.1}{25}$

or, $N_1 = 0.08$

∴ Normality of Fe^{+2} ion = 0.08

∴ Gram/litre of Fe^{2+} ion = Normality × Eq. wt.

= 0.08 × 55.85

= 4.47

∵ 1000 c.c. solution contains 4.47 g of iron

∴ 250 c.c. solution contains $\dfrac{4.47 \times 250}{1000}$ g of iron

= 1.12 g of iron

∵ 2 g of haematite contains 1.12 g of iron

∴ 100 g of haematite contains $\dfrac{1.12 \times 100}{2}$ g of iron

= 56 g of iron

Therefore, amount of iron in haematite (sample) = 56%.

Things to Remember

7 g of mercuric chloride is dissolved in 100 c.c. of water to make saturated solution. It is kept overnight, then the supernatant liquid is used.

14.15. To estimate the amount of ferrous and ferric ions in the given solution using 0.1N $K_2Cr_2O_7$

Theory: Ferrous ion of the mixture solution is directly estimated by titrating it with 0.1N $K_2Cr_2O_7$.

$$\begin{array}{l} Cr_2O_7^{2-} + 14H^+ + \cancel{6e} \rightarrow 2Cr^{3+} + 7H_2O \\ \left[Fe^{2+} \rightarrow Fe^{3+} + \cancel{e}\right] \times 6 \\ \hline Cr_2O_7^{2-} + 6Fe^{2+} + 14H^+ \rightarrow 2Cr^{3+} + 6Fe^{3+} + 7H_2O \end{array}$$

For estimation of ferric ion present in the mixture solution, ferric ion is first reduced to ferrous ion by $SnCl_2$.

$$\begin{array}{l} \left[Fe^{3+} + \cancel{e} \rightarrow Fe^{2+}\right] \times 2 \\ Sn^{2+} \rightarrow Sn^{4+} + 2\cancel{e} \\ \hline 2Fe^{3+} + Sn^{2+} \rightarrow 2Fe^{2+} + Sn^{4+} \end{array}$$

Excess of $SnCl_2$ must be removed completely as it is oxidised by $K_2Cr_2O_7$. This is done by $HgCl_2$ which reacts with $SnCl_2$ as shown below:

$$\begin{array}{l} Sn^{2+} \rightarrow Sn^{4+} + 2\cancel{e} \\ 2Hg^{2+} + 2\cancel{e} \rightarrow Hg_2^{2+} \\ \hline Sn^{2+} + 2Hg^{2+} \rightarrow Sn^{4+} + Hg_2^{2+} \end{array}$$

Now the total ferrous ion (original + reduced) is estimated by titrating with 0.1N $K_2Cr_2O_7$.

Procedure: It involves following steps:

(i) **First titration:** The burette is rinsed with 0.1N $K_2Cr_2O_7$ and filled with the same. Initial reading of the burette is noted. Now 25 c.c. of mixture solution is pipetted out and taken into the conical flask. Two test tubes of dilute sulphuric acid and 5 c.c. syrupy phosphoric acid are added to the flask. Now it is titrated against 0.1N $K_2Cr_2O_7$ using 8 drops of diphenylamine or sodium diphenylamine sulphonate. At the end point, the solution becomes bluish-violet. Final reading of the burette is noted.

The process is repeated till two concordant readings are obtained. From this titration ferrous ion can be estimated.

(ii) Second titration: 25 c.c. of the mixture solution is pipetted out and taken into a conical flask. 10 c.c. concentrated hydrochloric acid is added to it and heated to boil. Now stannous chloride solution is added dropwise from a burette till the yellow colour of the ferric ion turns to faint green (due to Fe^{2+} ions). The flask is then cooled under tap water. Now 100 c.c. of distilled water is added to it followed by 10 c.c. of $HgCl_2$ solution. On addition of $HgCl_2$ solution, white silky precipitate of Hg_2Cl_2 appears. This slight amount of mercurous ion has no appreciable effect on oxidising agent like $K_2Cr_2O_7$. Now 8 drops of diphenylamine or sodium diphenylamine sulphonate and 5 c.c. of syrupy phosphoric acid are added and then titrated with 0.1N $K_2Cr_2O_7$. At the end point, the solution becomes bluish-violet. The process is repeated till two concordant readings are obtained. From this titration, whole of the ferrous ion (original + reduced) can be estimated.

Observation and Calculation:

Table 30

Sl. No.	Vol. of mix. soln. (Fe^{2+} and Fe^{3+}) (c.c.)	Vol. of 0.1N $K_2Cr_2O_7$ taken in burette (c.c.)			Concordant reading (c.c.) (say)	Indicator	End point
		Initial reading (say)	Final reading (say)	Diff. (say)			
1.	25	6.00	19.80	13.80			
2.	25	5.40	18.75	13.35	13.35	Diphenyl-amine	Bluish-violet
3.	25	2.70	16.05	13.35			

25 c.c. Fe^{2+} ion (original) ≡ 13.35 c.c. 0.1N $K_2Cr_2O_7$

or, $V_1N_1 = V_2N_2$

or, $25 \times N_1 = 13.35 \times 0.1$

or, $N_1 = \dfrac{13.35 \times 0.1}{25}$

or, $N_1 = 0.053$

∴ Normality of Fe^{2+} ion = 0.053 ... (1)

∴ Gram/litre of Fe^{2+} ion

= Normality × Eq. wt.

$= 0.053 \times 55.85$

$= 2.96$

Table 31

Sl. No.	Vol. of mix. of soln. (original Fe^{2+} and reduced Fe^{2+}) (c.c.)	Vol. of 0.1N $K_2Cr_2O_7$ taken in burette (c.c.)			Concordant reading (c.c.) (say)	Indicator	End point
		Initial reading (say)	Final reading (say)	Diff. (say)			
1.	25	0.30	28.20	27.90			
2.	25	1.50	29.30	27.80	27.80	Diphenyl-amine	Bluish-violet
3.	25	2.40	30.20	27.80			

25 c.c. Fe^{2+} ion (original + reduced)

$\equiv 27.80$ c.c. 0.1N $K_2Cr_2O_7$

or, $V_1N_1 = V_2N_2$

or, $25 \times N_1 = 27.80 \times 0.1$

or, $N_1 = \dfrac{27.80 \times 0.1}{25}$

or, $N_1 = 0.111$

$\therefore$ Normality of Fe^{2+} ion (original + reduced)

$= 0.111$ ... (2)

$\therefore$ Normality of Fe^{3+} ion

= Normality of Fe^{2+} ion (original + reduced) – normality of Fe^{2+} ion (original)

$= 0.111 - 0.053$

$= 0.058$

$\therefore$ Gram/litre of Fe^{3+} ion = Normality × Eq. Wt.

$= 0.058 \times 55.85$

$= 3.24$

15. IODOMETRY AND IODIMETRY

15.1. To prepare 0.1N sodium thiosulphate solution and to standardise it by iodometric method using 0.1N $K_2Cr_2O_7$

Theory: Mol. wt. of $Na_2S_2O_3 \cdot 5H_2O$

$$= 23 \times 2 + 32 \times 2 + 16 \times 3 + 18 \times 5$$
$$= 46 + 64 + 48 + 90$$
$$= 248$$

The oxidation reaction of sodium thiosulphate may be represented as

$$\underset{\text{(thiosulphate)}}{2S_2O_3^{2-}} \rightarrow \underset{\text{(tetrathionate)}}{S_4O_6^{2-}} + 2e$$

$\because$ $2S_2O_3^{2-} \equiv 2e$

$\therefore$ $S_2O_3^{2-} \equiv 1e$

$\therefore$ Eq. wt. of sod. thiosulphate $= \dfrac{\text{Mol. wt.}}{1}$

$= 248$

$\because$ For 1000 c.c. N solution, 248 g of sod. thiosulphate is needed.

$\therefore$ For 1000 c.c. 0.1N solution, 248×0.1 g of sod. thiosulphate is needed.

$\therefore$ For 250 c.c. 0.1N solution, $\dfrac{248 \times 0.1 \times 250}{1000}$ g of sod. thiosulphate is needed.

= 6.2 g of sod. thiosulphate is needed

Thus 6.2 g of sodium thiosulphate is dissolved in water and the volume is made 250 c.c. This solution is now roughly 0.1N sodium thiosulphate solution.

Now potassium dichromate solution is treated with excess of KI and the liberated iodine is titrated with 0.1N (f = x) sodium thiosulphate solution.

$$Cr_2O_7^{2-} + 14H^+ + 6\cancel{e} \rightarrow 2Cr^{3+} + 7H_2O$$

$$[2I^- \rightarrow I_2 + 2\cancel{e}] \times 3$$

$$Cr_2O_7^{2-} + 6I^- + 14H^+ \rightarrow 2Cr^{3+} + 3I_2 + 7H_2O$$

$$\therefore \quad Cr_2O_7^{2-} \equiv 6I^- \equiv 3I_2$$

$$\therefore \quad \frac{Cr_2O_7^{2-}}{6} \equiv I^- \equiv I \qquad \text{... (1)}$$

$$2S_2O_3^{2-} \rightarrow S_4O_6^{2-} + 2\cancel{e}$$

$$I_2 + 2\cancel{e} \rightarrow 2I^-$$

$$2S_2O_3^{2-} + I_2 \rightarrow S_4O_6^{2-} + 2I^-$$

$$\therefore \quad 2S_2O_3^{2-} \equiv 2I^- \equiv I_2$$

$$\therefore \quad S_2O_3^{2-} \equiv I^- = I \qquad \text{... (2)}$$

From equations (1) and (2)

$$\frac{Cr_2O_7^{2-}}{6} \equiv I^- \equiv I \equiv S_2O_3^{2-} \qquad \text{... (3)}$$

Procedure: It involves following steps:

(i) **Preparation of 0.1N sodium thiosulphate:** First of all 6.2 g of sodium thiosulphate ($Na_2S_2O_3 \cdot 5H_2O$) is weighed and taken into a measuring flask. It is dissolved in distilled water and the volume is made 250 c.c. This solution is roughly 0.1N solution and hence it can be written as 0.1N (f = x) sodium thiosulphate solution.

(ii) **Titration:** 25 c.c. of 0.1N $K_2Cr_2O_7$ is taken in an iodine flask. A test tube of dilute sulphuric acid is added to it followed by 3 g of potassium iodide. Then 50 c.c. of distilled water is added to it and the mouth of the flask is closed with glass stopper. It is kept in dark for 5 minutes so that the liberation of iodine is complete. Now 0.1N (f = x) sodium thiosulphate is taken in a burette. Initial reading of the burette is noted. The flask is then kept below the nozzle of the burette and its solution is run down. When brown solution of the flask becomes light yellow, 2–3 c.c. of starch solution is added making the solution blue. The blue colour disappears at the end point. Final reading of the burette is noted. The process is repeated till two concordant readings are obtained.

Observation and Calculation:

First of all 6.2 g of hydrated sodium thiosulphate is dissolved in distilled water and the volume is made 250 c.c. It is now 0.1N (f = x) sodium thiosulphate solution.

25 c.c. of 0.1N $K_2Cr_2O_7$ is treated with KI and then liberated iodine is titrated against 0.1N (f = x) sodium thiosulphate.

Table 32

Sl. No.	Vol. of 0.1N $K_2Cr_2O_7$ (c.c.)	Vol. of 0.1N (f = x) sodium thiosulphate taken in burette (c.c.)			Concordant reading (c.c.) (say)	Indicator	End point
		Initial reading (say)	Final reading (say)	Diff. (say)			
1.	25	0.0	24.1	24.1			
2.	25	24.1	48.1	24.0	24.0	Starch	Discharge of blue colour
3.	25	2.3	26.3	24.0			

$$24.0 \text{ c.c. } 0.1\text{N } (f = x) \text{ sod. thiosulphate} \equiv 25 \text{ c.c. } 0.1\text{N } K_2Cr_2O_7$$

or, $V_1N_1 = V_2N_2$

or, $24 \times 0.1 \times x = 25 \times 0.1$

or, $x = \dfrac{25 \times 0.1}{24 \times 0.1}$

or, $x = 1.04$

$\therefore$ Normality of sod. thiosulphate = 0.1N (f = 1.04)

Things to Remember

(i) Sodium thiosulphate is not a primary standard because it is a hydrated substance and its water content is not constant. It is standardised by $K_2Cr_2O_7$.

(ii) Sod. thiosulphate is a reducing agent.

15.2. To determine the strength of given sodium thiosulphate solution by iodometric method using 0.1N $K_2Cr_2O_7$

Theory: Potassium dichromate solution is treated with excess potassium iodide and the liberated iodine is titrated with sodium thiosulphate solution.

$$Cr_2O_7^{2-} + 14H^+ + 6\not{e} \rightarrow 2Cr^{3+} + 7H_2O$$

$$[2I^- \rightarrow I_2 + 2\not{e}] \times 3$$

$$Cr_2O_7^{2-} + 6I^- + 14H^+ \rightarrow 2Cr^{3+} + 3I_2 + 7H_2O$$

$$\therefore \quad Cr_2O_7^{2-} \equiv 6I^- \equiv 3I_2$$

$$\therefore \quad \frac{Cr_2O_7^{2-}}{6} \equiv I^- \equiv I \qquad \text{... (1)}$$

$$2S_2O_3^{2-} \rightarrow S_4O_6^{2-} + 2\not{e}$$

$$I_2 + 2\not{e} \rightarrow 2I^-$$

$$2S_2O_3^{2-} + I_2 \rightarrow S_4O_6^{2-} + 2I^-$$

$$\therefore \quad 2S_2O_3^{2-} \equiv 2I^- \equiv I_2$$

$$\therefore \quad S_2O_3^{2-} \equiv I^- \equiv I \qquad \text{... (2)}$$

From equations (1) and (2)

$$\frac{Cr_2O_7^{2-}}{6} \equiv I^- \equiv I \equiv S_2O_3^{2-} \qquad \text{... (3)}$$

Procedure: 25 c.c. of 0.1N $K_2Cr_2O_7$ is taken in an iodine flask. A test tube of dilute sulphuric acid is added to it followed by 3 g of potassium iodide. Then 50 c.c. of distilled water is added to it and the mouth of the flask is closed with glass stopper. It is kept in dark for 5 minutes so that the liberation of iodine is complete. Now the given sodium thiosulphate solution is taken in a burette. Initial reading of the burette and is noted. The flask is then kept below the nozzle of the burette and its solution is run down. When brown solution of the flask becomes light yellow, 2–3 c.c. of starch is added. On addition of starch, blue colour appears which disappears at the end point. Final reading of the burette is noted. The process is repeated till two concordant readings are obtained.

Observation and Calculation:

First of all 25 c.c. of 0.1N $K_2Cr_2O_7$ is treated with KI and then liberated iodine is titrated against the given sodium thiosulphate solution.

Table 33

Sl. No.	Vol. of 0.1N $K_2Cr_2O_7$ (c.c.)	Vol. of sodium thiosulphate solution (given) taken in burette (c.c.)			Concordant reading (c.c.) (say)	Indicator	End point
		Initial reading (say)	Final reading (say)	Diff. (say)			
1.	25	0.0	24.1	24.1			
2.	25	24.1	48.1	24.0	24.0	Starch	Discharge of blue colour
3.	25	2.3	26.3	24.0			

24 c.c. sodium thiosulphate ≡ 25 c.c. 0.1N $K_2Cr_2O_7$

or, $V_1N_1 = V_2N_2$

or, $24 \times N_1 = 25 \times 0.1$

or, $N_1 = \dfrac{25 \times 0.1}{24}$

or, $N_1 = 0.104$

∴ Normality of sod. thiosulphate = 0.104

∴ Gram/litre of sod. thiosulphate

= Normality × Eq. wt.

= 0.104 × 248

= 25.8

15.3. To determine the percentage purity of blue vitriol ($CuSO_4 \cdot 5H_2O$) by iodometric method using 0.1N sodium thiosulphate

Theory: When potassium iodide is added to the copper sulphate solution, iodine is liberated which is then titrated against standard sodium thiosulphate solution using starch as indicator.

$$\left[Cu^{2+} + \cancel{e} \rightarrow Cu^{+}\right] \times 2$$

$$\frac{2I^{-} \rightarrow I_2 + 2\cancel{e}}{2Cu^{2+} + 2I^{-} \rightarrow 2Cu^{+} + I_2}$$

$$\therefore \quad Cu^{2+} \equiv I^{-} \equiv I \qquad \text{... (1)}$$

$$2S_2O_3^{2-} \rightarrow S_4O_6^{2-} + 2\cancel{e}$$

$$\frac{I_2 + 2\cancel{e} \rightarrow 2I^{-}}{2S_2O_3^{2-} + I_2 \rightarrow S_4O_6^{2-} + 2I^{-}}$$

$$\therefore \quad S_2O_3^{2-} \equiv I^{-} \equiv I \qquad \text{... (2)}$$

From equations (1) and (2)

$$Cu^{2+} \equiv S_2O_3^{2-} \equiv I^{-} \equiv I \qquad \text{... (3)}$$

The equation (3) shows that equivalent weights of blue vitriol and sodium thiosulphate are equal to their molecular weights.

$\therefore$ Eq. wt. of $CuSO_4{\cdot}5H_2O$

$= $ Mol. wt. of $CuSO_4{\cdot}5H_2O$

$= 63.5 + 32 + 16 \times 4 + 18 \times 5$

$= 63.5 + 32 + 64 + 90$

$= 249.5$

Procedure: It involves following steps:

(i) Preparation of solution of blue vitriol: First of all 3 g of blue vitriol is weighed on the watch glass and transferred into a measuring flask quantitatively. It is dissolved in distilled water and the volume made 250 c.c. The solution is made homogeneous.

(ii) Titration: Now 25 c.c. of solution of blue vitriol prepared as above is taken into an iodine flask with the help of pipette. To neutralise the solution, sodium carbonate solution is added dropwise till a slight turbidity appears. To dissolve the turbidity, dilute acetic acid is added. Thus a clear light blue solution is obtained. Then 3 g of potassium iodide is added. The mouth of the iodine flask is closed by glass stopper and it is kept in dark for 5 minutes. Now 0.1N sodium thiosulphate solution is taken in the burette. Initial reading of the burette is noted. Then solution of 0.1N sodium thiosulphate is run from the burette into the flask till the solution becomes light yellow. Now 2–3 c.c. of starch is added and again sodium thiosulphate is added. At the end point, the blue colour due to starch disappears. Final reading of the burette is noted. The process is repeated till two concordant readings are obtained.

Observation and Calculation:

First of all 3 g of blue vitriol is weighed and dissolved in water. The volume is made 250 c.c. Then 25 c.c. of this solution is treated with KI and the iodine liberated is titrated against 0.1N sodium thiosulphate using starch as indicator.

Table 34

Sl. No.	Vol. of blue vitriol soln. (c.c.)	Vol. of 0.1N sodium thiosulphate solution (given) taken in burette (c.c.)			Concordant reading (c.c.) (say)	Indicator	End point
		Initial reading (say)	Final reading (say)	Diff. (say)			
1.	25	4.10	16.20	12.10			
2.	25	16.20	28.20	12.00	12.00	Starch	Discharge of blue colour
3.	25	28.20	40.20	12.00			

25 c.c. blue vitriol ≡ 12 c.c. 0.1N sodium thiosulphate

or, $V_1N_1 = V_2N_2$

or, $25 \times N_1 = 12 \times 0.1$

or, $N_1 = \dfrac{12 \times 0.1}{25}$

or, $N_1 = 0.048$

∴ Normality of blue vitriol = 0.048

∴ Gram/litre of blue vitriol

= Normality × Eq. wt.

= 0.048 × 249.5

= 11.98

$\because$ 1000 c.c. solution contains 11.98 g of blue vitriol

$\therefore$ 250 c.c. solution contains $\frac{11.98 \times 250}{1000}$ g blue vitriol

= 2.995 g blue vitriol

$\because$ 3 g of supplied blue vitriol contains 2.995 g blue vitriol

$\therefore$ 100 g of supplied blue vitriol contains $\frac{2.995 \times 100}{3}$

= 99.8 g blue vitriol

Therefore, amount of blue vitriol = 99.8%.

Things to Remember

Before the addition of potassium iodide, the solution of blue vitriol is made neutral by sodium carbonate solution and acetic acid solution. In this condition, reaction of blue vitriol with potassium iodide is quantitative.

15.4. To estimate manganese dioxide in pyrolusite by iodometric method using 0.1N sodium thiosulphate

Theory: A weighed amount of pyrolusite is heated with excess of conc. HCl. The liberated chlorine gas is passed through KI solution and then the liberated iodine is titrated with standard solution of sodium thiosulphate.

$$MnO_2 + 4H^+ + 2\not{e} \rightarrow Mn^{2+} + 2H_2O$$

$$2Cl^- \rightarrow Cl_2 + 2\not{e}$$

$$MnO_2 + 2Cl^- + 4H^+ \rightarrow Mn^{2+} + 2H_2O + Cl_2$$

$\therefore$ $MnO_2 \equiv Cl_2$ ... (1)

$$\therefore \text{ Eq. wt. of } MnO_2 = \frac{\text{Mol. wt.}}{2} = \frac{55 + 16 \times 2}{2} = \frac{55 + 32}{2} = \frac{87}{2} = 43.5$$

$$\begin{array}{l} Cl_2 + 2\cancel{e} \rightarrow 2Cl^- \\ 2I^- \rightarrow I_2 + 2\cancel{e} \\ \hline Cl_2 + 2I^- \rightarrow 2Cl^- + I_2 \end{array}$$

$$\therefore \quad Cl_2 \equiv I_2 \qquad \ldots (2)$$

$$\begin{array}{l} 2S_2O_3^{2-} \rightarrow S_4O_6^{2-} + 2\cancel{e} \\ I_2 + 2\cancel{e} \rightarrow 2I^- \\ \hline 2S_2O_3^{2-} + I_2 \rightarrow S_4O_6^{2-} + 2I^- \end{array}$$

$$\therefore \quad I_2 \equiv 2S_2O_3^{2-} \qquad \ldots (3)$$

From equations (1), (2) and (3)

$$MnO_2 \equiv Cl_2 \equiv I_2 \equiv 2S_2O_3^{2-}$$

or, $$\frac{MnO_2}{2} = Cl \equiv I \equiv S_2O_3^{2-} \qquad \ldots (4)$$

Procedure: It involves following steps:

(i) Preparation of solution: First of all 1 g of pyrolusite is weighed and taken in a distillation flask. Two test tubes of concentrated hydrochloric acid are added to it. The flask is heated and the chlorine gas evolved is passed into 100 c.c. of 10% potassium iodide. When the total chlorine is passed, the iodine solution so obtained is transferred into a 250 c.c. measuring flask. The solution is diluted with distilled water to make the volume 250 c.c. The solution is made homogeneous by shaking.

(ii) Titration: The burette is rinsed with 0.1N sodium thiosulphate and filled by it. Initial reading of the burette is noted. Now 25 c.c. of the iodine solution prepared as above is taken in a conical flask. Then 0.1N sodium thiosulphate solution is run from the burette into the conical flask until the solution becomes light yellow. Now 2–3 c.c. of starch is added and again sodium thiosulphate is added till the blue colour of the solution disappears. The process is repeated till two concordant readings are obtained.

Observation and Calculation:

Wt. of pyrolusite = 1 g

Eq. wt. of MnO_2 = 43.5

Table 35

Sl. No.	Vol. of iodine soln. produced by MnO_2 (c.c.)	Vol. of 0.1N sodium thiosulphate taken in burette (c.c.)			Concordant reading (c.c.) (say)	Indicator	End point
		Initial reading (say)	Final reading (say)	Diff. (say)			
1.	25	1.20	21.30	20.10			
2.	25	1.60	21.60	20.00	20.00	Starch	Discharge of blue colour
3.	25	4.80	24.80	20.00			

25 c.c. MnO_2 solution ≡ 20 c.c. 0.1N sod. thiosulphate

or, $V_1N_1 = V_2N_2$

or, $25 \times N_1 = 20 \times 0.1$

or, $N_1 = \dfrac{20 \times 0.1}{25}$

or, $N_1 = 0.08$

∴ Normality of MnO_2 solution = 0.08

∴ Gram/litre of MnO_2 solution

= Normality × Eq. wt.

= 0.08 × 43.5

= 3.48

∵ 1000 c.c. solution contains 3.48 g MnO_2

∴ 250 c.c. solution contains $\dfrac{3.48 \times 250}{1000}$ g MnO_2

= 0.87 g of MnO_2

∵ 1 g of pyrolusite contains = 0.87 g of MnO_2

∴ 100 g of pyrolusite contains 0.87 × 100 g of MnO_2

= 87 g of MnO_2

Therefore, amount of MnO_2 in pyrolusite = 87%.

15.5. To determine the normality of the given iodine solution using 0.1N sodium thiosulphate solution

Theory: The given solution of iodine is titrated with 0.1N sodium thiosulphate using starch as indicator.

$$2S_2O_3^{2-} \rightarrow S_4O_6^{2-} + 2\cancel{e}$$

$$I_2 + 2\cancel{e} \rightarrow 2I^-$$

$$\overline{2S_2O_3^{2-} + I_2 \rightarrow S_4O_6^{2-} + 2I^-}$$

Procedure: The burette is rinsed with 0.1N sodium thiosulphate and filled by it. Initial reading of the burette is noted. Now 25 c.c. of iodine solution (prepared as described later) is pipetted out and taken into a conical flask. The sodium thiosulphate solution is run from the burette into the conical flask till the solution of the flask becomes light yellow. Now 2–3 c.c. of starch is added and again thiosulphate is added. At the end point, the blue colour of the solution due to starch disappears. Final reading of the burette is noted. The whole process is repeated till two concordant readings are obtained.

Observation and Calculation:

Table 36

Sl. No.	Vol. of iodine soln. (c.c.)	Vol. of 0.1N sodium thiosulphate (c.c.)			Concordant reading (c.c.) (say)	Indicator	End point
		Initial reading (say)	Final reading (say)	Diff. (say)			
1.	25	0.0	24.8	24.8			
2.	25	24.8	49.5	24.7	24.7	Starch	Discharge of blue colour
3.	25	2.0	26.7	24.7			

25 c.c. iodine solution

$\equiv$ 24.7 c.c. 0.1N sod. thiosulphate

or, $V_1N_1 = V_2N_2$

or, $25 \times N_1 = 24.7 \times 0.1$

or, $N_1 = \dfrac{24.7 \times 0.1}{25}$

or, $N_1 = 0.099$

$\therefore$ Normality of iodine solution = 0.099.

Things to Remember

(i) Iodine is available in a pure form. Due to its volatility, its exact weighing is difficult. It is also very slightly soluble in water, but is readily soluble in aqueous solution of potassium iodide due to formation of brown coloured ion.

$$I_2 + I^- \rightarrow I_3^-$$

or, $I_2 + KI \rightarrow KI_3$

Potassium tri-iodide readily liberates iodine during the titration with reducing agents. Hence in volumetric analysis, an iodine solution is prepared by dissolving iodine in potassium iodide solution.

(ii) Iodine should be weighed in weighing bottle because it is volatile.

(iii) Reduction of iodine can be shown as under:

$$I_2 + 2e \rightarrow 2I^-$$

$\therefore$ $I_2 \equiv 2e$

$\therefore$ $I \equiv e$

$\therefore$ Eq. wt. of iodine = at. wt. of iodine

= 127

(iv) For preparation of 0.1N solution of iodine, 12.7 g of it is dissolved in aqueous solution of 20 g of potassium iodide and the volume is made 1000 c.c.

(v) The above experiment is an example of iodimetry because direct iodine solution is being titrated.

(vi) The above experiment is actually standardisation of iodine solution.

16. PRECIPITATION TITRATION

16.1. To estimate the amount of sodium chloride per litre in the supplied solution using 0.05N $AgNO_3$ by Mohr's method

Theory: The supplied solution of sodium chloride is titrated with standard solution of silver nitrate using 5% potassium chromate as an indicator (precipitation indicator). The reaction taking place is quantitative.

$$\underset{\substack{108+14+48 \\ =170}}{AgNO_3} + \underset{\substack{23+35.5 \\ =58.5}}{NaCl} \rightarrow AgCl + NaNO_3$$

$$\text{Eq. wt. of NaCl} = \text{Molecular mass of NaCl} = 58.5$$

Procedure: The burette is rinsed with 0.05 N $AgNO_3$ and filled by it. Initial reading of the burette is noted. Now 25 c.c. of the supplied NaCl solution is pipetted out and taken into a conical flask. 5 drops of precipitation indicator (5% K_2CrO_4) are added. Then silver nitrate solution is run from the burette into the conical flask. At the end point light brick red coloured precipitate is obtained. Final reading of the burette is noted. The process is repeated till two concordant readings are obtained.

Observation and Calculation:

Table 37

Sl. No.	Vol. of NaCl soln. (c.c.)	Vol. of 0.05N $AgNO_3$ taken in burette (c.c.)			Concordant reading (c.c.) (say)	Indicator	End point
		Initial reading (say)	Final reading (say)	Diff. (say)			
1.	25	3.0	27.5	24.5			
2.	25	6.3	30.7	24.4	24.4	K_2CrO_4	Brick red precipitate
3.	25	3.6	28.0	24.4			

25 c.c. NaCl solution ≡ 24.4 c.c. 0.05N $AgNO_3$

or, $V_1N_1 = V_2N_2$

or, $25 \times N_1 = 24.4 \times 0.05$

or, $N_1 = \frac{24.4 \times 0.05}{25}$

or, $N_1 = 0.049$

$\therefore$ Normality of NaCl = 0.049

$\therefore$ Gram/litre of NaCl = Normality × Eq. wt.

= 0.049 × 58.5

= 2.87

Things to Remember

(i) Estimation of amount of sodium chloride in the solution can also be done by Fajan's method. In this method fluorescein (adsorption indicator) is used instead of potassium chromate. Rest process is same as in the Mohr's method.

(ii) All the solutions should be prepared in distilled water and titration flask should also be rinsed with distilled water before every titration.

(iii) Pinch-cock type burette should not be used as silver nitrate attacks rubber.

(iv) In Mohr's and Fajan's methods, titrations are carried out in cold and neutral condition.

16.2. To determine the strength of a given potassium thiocyanate solution using 0.05N $AgNO_3$ by Volhard's method

Theory: Silver nitrate solution is titrated with potassium thiocyanate solution using ferric alum (ferric ammonium sulphate) as an indicator. The reaction taking place in the titration can be shown as under:

$$Ag^{+} + \overline{\ddot{S}}CN \rightarrow \underset{\text{(White)}}{AgSCN} \downarrow$$

Procedure: The burette is rinsed with potassium thiocyanate solution and filled by it. Initial reading of the burette is noted. Now 25 c.c. of 0.05N $AgNO_3$ is pipetted out and taken into a conical flask. To it 1 c.c. of ferric alum indicator is added.

The flask is kept below the nozzle of the burette and the sodium thiocyanate solution is run from the burette into the flask. The flask is stirred vigorously during the titration. At the end point brownish red precipitate is obtained. Final reading of the burette is noted.

Observation and Calculation:

Table 38

Sl. No.	Vol. of 0.05N $AgNO_3$ (c.c.)	Vol. of KSCN solution taken in burette (c.c.)			Concordant reading (c.c.) (say)	Indicator	End point
		Initial reading (say)	Final reading (say)	Diff. (say)			
1.	25	3.2	27.3	24.1			
2.	25	2.1	26.1	24.0	24.0	Ferric alum	Brownish-red ppt.
3.	25	5.5	29.5	24.0			

$$24 \text{ c.c. KSCN} \equiv 25 \text{ c.c. } 0.05\text{N } AgNO_3$$

or, $V_1N_1 = V_2N_2$

or, $24 \times N_1 = 25 \times 0.05$

or, $N_1 = \dfrac{25 \times 0.05}{24}$

or, $N_1 = 0.052$

$\therefore$ Normality of KSCN = 0.052

$\therefore$ Gram/litre of KSCN

= Normality × Eq. wt.

= 0.052 × 97

= 5.044

Mol. wt. of KSCN

= 39 + 32 + 12 + 14

= 97

$\therefore$ Eq. wt. = 97

Things to Remember

(i) The Volhard's method is a good method as it can be carried out in the presence of acid.

(ii) Hydrochloric acid and nitric acid can be analysed from the solution of their mixture. Total amount of acid can be estimated using standard sodium hydroxide and hydrochloric acid can be estimated using standard $AgNO_3$. Before starting the second titration, the solution must be neutralised by $CaCO_3$ and then the chloride ion is titrated. Indicator used in this titration may be K_2CrO_4 or fluorescein.

(iii) Percentage composition of KCl and NaCl may be determined from their mixture. 5 g of the mixture may be dissolved in 1000 c.c. of distilled water.

Let wt. of KCl = a g

$\therefore$ Wt. of NaCl = (5 – a) g

$$\therefore \text{ Normality of KCl } = \frac{a}{74.5} \quad \text{... (a)}$$

$$\therefore \text{ Normality of NaCl } = \frac{5-a}{58.5} \quad \text{... (b)}$$

Normality of the mixture solution may be determined by titrating it with standard $AgNO_3$ using K_2CrO_4 or flourescein. Then sum of equations (a) and (b) will be equal to the normality of the mixture solution.

(iv) Percentage composition of NaCl and HCl from their mixture solution may be determined. The amount of HCl can be determined by titrating it with standard NaOH solution and the total amount of these can be estimated by standard $AgNO_3$ using K_2CrO_4 or fluorescein. Before the second titration, the mixture solution must be neutralised by $CaCO_3$.

17. COMPLEXOMETRIC TITRATION

17.1. To determine magnesium in given solution using 0.01M EDTA by direct titration

Theory: The solution of magnesium ion can be titrated with 0.01M disodium salt of EDTA using mixture of Eriochrome Black T (Solochrome Black) and potassium nitrate as indicator.

$$H_2Y^{2-} + Mg^{2+} \rightarrow MgY^{2-} + 2H^+$$

$$\text{or, } \underline{Na_2H_2Y} + \underline{MgSO_4} \rightarrow Na_2MgY + H_2SO_4$$

or, 372 g $Na_2H_2Y \equiv 24$ g Mg

or, $\frac{372}{100}\text{ g} \equiv \frac{24}{100}\text{ g}$

or, 3.72 g ≡ 0.24 g

Mol. formula of disodium EDTA $= C_{10}H_{14}O_8N_2Na_2 \cdot 2H_2O$

Mol. wt. = 372

or, 1000 c.c. of 0.01M EDTA ≡ 0.24 g of Mg

= 240 mg of Mg

or, 1 c.c. of 0.01M EDTA ≡ 0.24 mg of Mg

Procedure: The burette is rinsed with 0.01M EDTA and filled by it. Initial reading of the burette is noted. 25 c.c. of the given solution of magnesium sulphate is pipetted out and taken into a conical flask. 25 c.c. of distilled water is added to it. Then 2 c.c. of buffer solution (pH = 10) is added. Now 30–50 mg of a solid mixture of the indicator (EBT) with A.R. potassium nitrate (mixture should contain 1% of EBT) is added and heated to 40°C till the solution becomes wine red. 0.01M EDTA is then run from the burette into the flask. The conical flask is vigorously and constantly shaken during the titration. At the end point wine red colour of the solution changes to blue. Final reading of the burette is noted. The process is repeated till two concordant readings are obtained.

Observation and Calculation:

Table 39

Sl. No.	Vol. of magnesium sulphate solution (c.c.)	Vol. of 0.01M EDTA taken in burette (c.c.)			Concordant reading (c.c.) (say)	Indicator	End point
		Initial reading (say)	Final reading (say)	Diff. (say)			
1.	25	0.0	25.1	25.1			
2.	25	5.0	30.0	25.0	25.0	EBT	Blue
3.	25	3.2	28.2	25.0			

∵ 1 c.c. 0.01M EDTA ≡ 0.24 mg of Mg

∴ 25 c.c. of 0.01M EDTA ≡ (0.24 × 25) mg of Mg

≡ 6 mg of Mg ... (1)

From titration,

25 c.c. of mag. sulphate

$\equiv$ 25 c.c. of 0.01M EDTA ... (2)

From equations (1) and (2),

25 c.c. of mag. sulphate $\equiv$ 6 mg of Mg

$$\therefore \quad 1000 \text{ c.c. of mag. sulphate} = \frac{6}{25} \times 1000 \text{ mg of Mg}$$

$$= 240 \text{ mg of Mg}$$

$$= 0.24 \text{ g of Mg}$$

Things to Remember

(i) Preparation of 0.01M EDTA solution: EDTA is an abbreviation for Ethylene Diamine Tetraacetic Acid. But in complexometric titration, its disodium salt is always used as it is readily soluble in water. Hence EDTA written anywhere implies disodium salt.

Mol. formula of EDTA = $C_{10}H_{14}O_8N_2Na_2 \cdot 2H_2O$

Mol. wt. of EDTA = 372

$\because$ 1000 c.c. M solution needs 372 g of EDTA

$\therefore$ 1000 c.c. 0.01M solution needs 372 × 0.01 g of EDTA = 3.72 g of EDTA

Thus 3.72 g of EDTA is dissolved in distilled water and the volume is made 1000 c.c. to make its strength 0.01M.

(ii) Preparation of magnesium sulphate solution: Most titrations are successful with 0.25 millimole of the metal ion present in a volume of 50–150 c.c. of solution. If the metal ion concentration is too high, then the end point may be very difficult to discern. Hence it is advisable that the strength of magnesium sulphate solution be 0.01M.

Mol. formula of mag. sulphate = $MgSO_4 \cdot 7H_2O$

Mol. wt. of mag. sulphate = 246

$\because$ 1000 c.c. of M solution needs 246 g of $MgSO_4 \cdot 7H_2O$

∴ 1000 c.c. of 0.01M solution needs 246×0.01 g of $MgSO_4 \cdot 7H_2O$ = 2.46 g of $MgSO_4 \cdot 7H_2O$

Thus 2.46 g of hydrated magnesium sulphate is weighed and dissolved in distilled water. Then volume is made 1000 c.c. to make the strength 0.01M.

(iii) Preparation of buffer solution (pH = 10): For many EDTA titrations the pH of the solution is extremely critical; often limits of ±1 unit of pH, and frequently limits of ±0.5 unit of pH must be achieved for a successful titration to be carried out. To achieve such narrow limits of control it is necessary to make use of a pH meter.

For preparing buffer solution, 57 c.c. of concentrated ammonia solution is added to 7 g of A.R. ammonium chloride and diluted to 100 c.c. with distilled water.

(iv) Preparation of indicator: Eriochrome Black T is abbreviated as EBT. It is also known as Solochrome Black.

Solid mixture of the indicator EBT with A.R. potassium nitrate is prepared by mixing 1 mg of EBT with 99 mg of A.R. potassium nitrate. Only 30–50 mg of this mixture may be used as indicator at a time. Addition of excess amount of indicator is prohibited. If excess amount of indicator is used it may exhibit dicroism, i.e., there is an intermediate colour change one to two drops before the clear end point.

Solution of EBT may also be used here as an indicator. The solution of EBT is prepared by dissolving 0.2 g of EBT in 15 c.c. of triethanol amine and adding 5 c.c. of absolute alcohol to reduce the viscosity of the solution.

However, it should be noted that satisfactory colour at end point can be achieved by the use of 30–50 mg of solid mixture of EBT and KNO_3 in which EBT be 1%.

(v) Colour at end point: When EBT is added to the solution of Mg^{2+} ion, colour of the indicator changes from blue to red due to formation of indicator—magnesium complex. At the end point the indicator will be set free and colour change will be from red to blue.

The colour change near the end point is very slow. Hence care must be taken to avoid over titration.

(vi) Attainment of the end point: In many EDTA titrations the colour change in the neighbourhood of the end point may be slow. In such cases, cautious addition of the titrant coupled with continuous stirring of the solution is advisable; the use of magnetic stirrer is recommended. Generally, a sharper end point may be achieved if the solution is warmed to about 40°C.

17.2. To determine calcium in given solution using 0.01M EDTA by substitution titration

Theory: The solution of calcium ion can be titrated with 0.01M EDTA. The reaction of the titration may be shown as below:

$$H_2Y^{2-} + Ca^{2+} \rightarrow CaY^{2-} + 2H^+$$

or, $Na_2H_2Y + Ca^{2+} \rightarrow Na_2CaY + 2H^+$

or, $372 \text{ g } Na_2H_2Y \equiv 40 \text{ g Ca}$

or, $\frac{372}{100} \text{ g } Na_2H_2Y \equiv \frac{40}{100} \text{ g Ca}$

or, $3.72 \text{ g } Na_2H_2Y \equiv 0.4 \text{ g Ca}$

Mol. formula of disodium EDTA $= C_{10}H_{14}O_8N_2Na_2 \cdot 2H_2O$

Mol. wt. = 372

or, 1000 c.c. of 0.01M EDTA $\equiv$ 0.4 g of Ca

$\equiv$ 400 mg of Ca

or, 1 c.c. of 0.01M EDTA $\equiv$ 0.4 mg of Ca

Procedure: The burette is rinsed with 0.01M EDTA and filled by it. Initial reading of the burette is noted. Now 25 c.c. of the calcium ion solution is pipetted out and taken into a conical flask. It is diluted with 25 c.c. distilled water followed by the addition of 2 c.c. of buffer solution,

1 c.c. 0.1M Mg–EDTA, and 30–40 mg EBT/KNO_3 mixture containing 1% EBT. The solution is heated to 40°C and titrated with 0.01M EDTA. At the end point wine red colour of the solution changes to blue. (No tinge of reddish hue should remain at the equivalence point and titration should be slow at the end point.) Final reading of the burette is noted. The process is repeated till two concordant readings are obtained.

Observation and Calculation:

Table 40

Sl. No.	Vol. of Ca^{2+} ion solution (c.c.)	Vol. of 0.01M EDTA taken in burette (c.c.)			Concordant reading (c.c.) (say)	Indicator	End point
		Initial reading (say)	Final reading (say)	Diff. (say)			
1.	25	0.0	25.1	25.1			
2.	25	2.2	27.2	25.0	25.0	EBT	Blue
3.	25	5.0	30.0	25.0			

$\because$ 1 c.c. 0.01M EDTA $\equiv$ 0.4 mg of Ca

$\therefore$ 25 c.c. of 0.01M EDTA $\equiv$ (0.4 × 25) mg of Ca

$\equiv$ 10 mg of Ca ... (1)

From titration,

25 c.c. of 0.01M EDTA

$\equiv$ 25 c.c. of Ca^{2+} ion solution ... (2)

From equations (1) and (2),

25 c.c. of Ca^{2+} ion solution $\equiv$ 10 mg of Ca

$$\therefore \text{ 1000 c.c. of } Ca^{2+} \text{ ion solution} \equiv \frac{10}{25} \times 1000 \text{ mg of Ca}$$

$\equiv$ 400 mg of Ca

$\equiv$ 0.4 g of Ca

Things to Remember

(i) Discussion of experiment: When calcium ions are titrated with EDTA a relatively stable calcium complex is formed.

$$Ca^{2+} + H_2Y^{2-} \rightarrow \underset{\text{(relatively stable)}}{CaY^{2-}} + 2H^+$$

With calcium ions alone, no sharp end point can be obtained with Solochrome Black indicator and the transition from red to pure blue is not observed. With magnesium ions, a somewhat less stable complex is formed.

$$Mg^{2+} + H_2Y^{2-} \rightarrow \underset{\text{(relatively less stable)}}{MgY^{2-}} + 2H^+$$

But the magnesium-indicator complex is more stable than calcium-indicator complex and less stable than magnesium-EDTA complex. Consequently, during the titration of a solution containing magnesium and calcium ions with EDTA in the presence of Solochrome Black the EDTA reacts first with free calcium ions, then with the free magnesium ions, and finally with the magnesium-indicator complex. Since the magnesium-indicator complex is wine red in colour and the free indicator is blue between pH 7 and 11 the colour of the solution changes from wine red to blue at the end point.

(ii) Preparation of 0.01M EDTA: As in section 17.1.

(iii) Preparation of 0.01M Ca^{2+} ion solution:

Molecular formula of calcium carbonate = $CaCO_3$

Molecular weight of calcium carbonate = 100

∵ For 1000 c.c. M solution 100 g of $CaCO_3$ is needed.

∴ For 1000 c.c. 0.01M solution 100 × 0.01 g of $CaCO_3$ is needed.

∴ For 100 c.c. 0.01M solution $\frac{100 \times 0.01 \times 100}{1000}$ g of $CaCO_3$ is needed = 0.1 g of $CaCO_3$ is needed.

0.1 g of A.R. sample of $CaCO_3$ is weighed and taken in a 100 c.c. measuring flask. 5N HCl is added dropwise till effervescence ceases and the solution is clear. Now a few drops of methyl red indicator are added and then 0.1N NaOH solution is added dropwise till the red colour is discharged. Lastly, the solution is diluted up to the mark. This is neutral 0.01M solution of Ca^{2+} ion.

(iv) Preparation of 0.1M Mg–EDTA complex: The desired complex is prepared by mixing equal volumes (25 c.c.) of 0.2M solutions of EDTA and magnesium sulphate. The solution is neutralised (pH = 8–9) by adding one drop of phenolphthalein and 0.1N NaOH dropwise till the solution becomes faint pink. Now 1–2 drops of buffer solution and 30 mg of Solochrome Black/ KNO_3 mixture are added to it. A violet colour should be produced which turns blue on the addition of a drop of 0.01M EDTA solution and red on addition of a single drop of 0.01M $MgSO_4$ solution. This confirms equimolarity of magnesium and EDTA. If the solution does not pass this test, it may be treated with more EDTA or with more magnesium sulphate solution until the required condition of equimolarity is attained. Now this solution is approximately 0.1M Mg-EDTA complex solution. Alternatively solid Mg-EDTA complex may be used, which is available from Hopkin and Williams Ltd.

(v) Preparation of buffer solution: As in section 17.1.

(vi) Preparation of indicator: As in section 17.1.

(vii) No reddish tinge should remain at the end point.

(viii) Titration should be done slowly near the end point so as to avoid over titration.

(ix) Vessels of ordinary (soda) glass should not be used for the storage of EDTA solution. In the course of time such soft glass containers will yield appreciable amounts of cations (including Ca^{2+} and Mg^{2+}) and anions to the solution of EDTA. Polythene bottles are the most satisfactory for the storage of EDTA solution.

17.3. To determine calcium in given solution using 0.01M EDTA by direct titration

Theory: Solution of calcium ion can be directly titrated with EDTA using murexide as indicator.

$$H_2Y^{2-} + Ca^{2+} \rightarrow CaY^{2-} + 2H^+$$

or, $Na_2H_2Y + Ca^{2+} \rightarrow Na_2CaY + 2H^+$

or, 372 g $Na_2H_2Y \equiv$ 40 g Ca

or, $\frac{372}{100}\text{ g} \equiv \frac{40}{100}\text{ g}$

or, 3.72 g ≡ 0.4 g

Mol. formula of disodium EDTA $= C_{10}H_{14}O_8N_2Na_2 \cdot 2H_2O$

Mol. wt. = 372

or, 1000 c.c. of 0.01M EDTA ≡ 0.4 g of Ca

≡ 400 mg of Ca

or, 1 c.c. of 0.01M EDTA ≡ 0.4 mg of Ca

Procedure: The burette is rinsed with 0.01M EDTA and filled by it. Initial reading of the burette is noted. Now 25 c.c. of the calcium ion solution is pipetted out and taken into a conical flask. It is diluted to 100 c.c. with distilled water. 2 c.c. of M NaOH solution is added to have a solution of pH about 12. Now 30–50 mg of murexide/KNO_3 mixture is added to the flask. Then 0.01M EDTA is run from the burette into the flask. At the end point colour changes from red to purple. Final reading of the burette is noted.

Observation and Calculation:

Table 41

Sl. No.	Vol. of Ca^{2+} ion solution (c.c.)	Vol. of 0.01M EDTA taken in burette (c.c.)			Concordant reading (c.c.) (say)	Indicator	End point
		Initial reading (say)	Final reading (say)	Diff. (say)			
1.	25	0.0	25.1	25.1			
2.	25	2.2	27.2	25.0	25.0	Murexide	Purple
3.	25	5.5	30.5	25.0			

∵ 1 c.c. of 0.01M EDTA ≡ 0.4 mg of Ca

$\therefore$ 25 c.c. of 0.01M EDTA $\equiv$ (0.4 × 25) mg of Ca

$\equiv$ 10 mg of Ca ... (1)

From titration,

25 c.c. of 0.01M EDTA

$\equiv$ 25 c.c. of Ca^{2+} ion solution ... (2)

From equations (1) and (2),

25 c.c. of Ca^{2+} ion solution $\equiv$ 10 mg of Ca

$\therefore$ 1000 c.c. of Ca^{2+} ion solution $\equiv \frac{10}{25} \times 1000$ mg of Ca

$\equiv$ 400 mg of Ca

$\equiv$ 0.4 g of Ca

Things to Remember

(i) Preparation of 0.01M EDTA: As in section 17.1.

(ii) Preparation of 0.01M calcium ion solution:

Molecular formula of calcium carbonate = $CaCO_3$

Molecular weight of calcium carbonate = 100

$\because$ For 1000 c.c. M $CaCO_3$ solution 100 g of $CaCO_3$ is needed.

$\therefore$ For 1000 c.c. 0.01M $CaCO_3$ solution 100 × 0.01 g of $CaCO_3$ is needed.

$\therefore$ For 100 c.c. 0.01M $CaCO_3$ solution $\frac{100 \times 0.01 \times 100}{1000}$ g of $CaCO_3$ is needed = 0.1 g of $CaCO_3$ is needed.

0.1 g of A.R. $CaCO_3$ is weighed and taken into a 100 c.c. measuring flask. It is dissolved in minimum quantity of 5M HCl adding drop by drop until effervescence stops and the solution becomes clear. Now the solution is diluted up to the mark.

(iii) Preparation of M NaOH solution: 4 g of NaOH is dissolved in distilled water and the volume is made 100 c.c. to get M solution.

(iv) Preparation of indicator: The indicator is prepared by grinding 0.1 g murexide with 10 g A.R. potassium nitrate.

17.4. To determine Calcium and Magnesium in a given solution using 0.01M EDTA

Theory: The solution is first titrated with 0.01M EDTA in presence of EBT/KNO_3 mixture. It gives total amount of calcium and magnesium. Only calcium can be estimated by titrating the solution with 0.01M EDTA in presence of murexide in strong alkaline medium. In such a medium magnesium gets precipitated and then only calcium is titrated.

1 c.c. of 0.01M EDTA ≡ 0.4 mg of Ca

1 c.c. of 0.01M EDTA ≡ 0.24 mg of Mg

Procedure: It involves following steps:

(i) First titration: The burette is rinsed with 0.01M EDTA and filled by it. Initial reading of the burette is noted. Now 25 c.c. of the solution is pipetted out and taken into a conical flask. 50 c.c. of distilled water is added followed by 4 c.c. of buffer solution (pH = 10) and 30–50 mg of EBT/KNO_3 mixture. The solution is heated to 60°. Then 0.01M EDTA is run from the burette into the flask. At the end point colour of the solution changes from wine red to blue. Final reading of the burette is noted. This titration gives total amount of calcium and magnesium.

(ii) Second titration: The burette is filled by 0.01M EDTA. Initial reading is noted. 25 c.c. of the solution (under examination) is taken into a conical flask. To this 5 c.c. of 2M NaOH is added followed by 30–50 mg of murexide/KNO_3 mixture and 25 c.c. of distilled water. Now the 0.01M EDTA is run from the burette into the flask. At the end point the colour of the solution changes from red to purple. Final reading of the burette is noted. This titration gives the amount of calcium as magnesium gets precipitated as magnesium hydroxide in strong alkaline medium.

Observation and Calculation:

Table 42: Estimation of Ca and Mg

Sl. No.	Vol. of solution containing Ca^{2+} and Mg^{2+} ions (c.c.)	Vol. of 0.01M EDTA taken in burette (c.c.)			Concordant reading (c.c.) (say)	Indicator	End point
		Initial reading (say)	Final reading (say)	Diff. (say)			
1.	25	0.0	25.1	25.1			
2.	25	2.3	27.3	25.0	25.0	EBT	Blue
3.	25	3.2	28.2	25.0			

Table 43: Estimation of Ca

Sl. No.	Vol. of solution containing Ca^{2+} ion (c.c.)	Vol. of 0.01M EDTA taken in burette (c.c.)			Concordant reading (c.c.) (say)	Indicator	End point
		Initial reading (say)	Final reading (say)	Diff. (say)			
1.	25	0.0	12.6	12.6			
2.	25	12.6	25.1	12.5	12.5	Murexide	Purple
3.	25	25.1	37.6	12.5			

25 c.c. of Ca^{2+} and Mg^{2+} ions

$\equiv$ 25 c.c. of 0.01M EDTA ... (1)

25 c.c. of Ca^{2+} ion

$\equiv$ 12.5 c.c. of 0.01M EDTA ... (2)

$\therefore$ 25 c.c. of Mg^{2+} ion

$\equiv$ (25 – 12.5) c.c. of 0.01M EDTA

= 12.5 c.c. of of 0.01M EDTA ... (3)

Estimation of Calcium:

$\because$ 1 c.c. of 0.01M EDTA $\equiv$ 0.4 mg of Ca

$\therefore$ 12.5 c.c. of 0.01M EDTA $\equiv$ (0.4 × 12.5) mg of Ca

= 5 mg of Ca

$\because$ 25 c.c. solution contains 5 mg of Ca

$\therefore$ 1000 c.c. solution contains $\frac{5 \times 1000}{25}$ mg of Ca

$$= \frac{5 \times 1000}{25 \times 1000} \text{ g of Ca}$$
$$= 0.2 \text{ g Ca}$$

Therefore, amount of calcium = 0.2 g/litre

Estimation of magnesium:

$\because$ 1 c.c. of 0.01M EDTA $\equiv$ 0.24 mg of Mg

$\therefore$ 12.5 c.c. of 0.01M EDTA $\equiv$ (0.24 × 12.5) mg of Mg

= 3 mg of Mg

$\because$ 25 c.c. solution contains 3 mg of Mg

$\therefore$ 1000 c.c. of solution contains $\frac{3 \times 1000}{25}$ mg of Mg

$$= \frac{3 \times 1000}{25 \times 1000} \text{ g of Mg}$$
$$= 0.12 \text{ g Mg}$$

Therefore, amount of magnesium = 0.12 g/litre

Things to Remember

(i) Preparation of 0.01 M EDTA: As in section 17.1.

(ii) Preparation of 0.01M Ca^{2+} and Mg^{2+} ions solution: One litre of this solution will contain 0.005 mole of $CaCO_3$ and 0.005 mole of $MgSO_4 \cdot 7H_2O$.

500 c.c. of $CaCO_3$ solution is made by dissolving its 0.005/2 mole in minimum volume of 5N HCl and diluting it to 500 c.c.

500 c.c. of $MgSO_4 \cdot 7H_2O$ solution is made by dissolving its 0.005/2 mole in distilled water and making the volume to 500 c.c.

When the two solutions are mixed together it will be 0.01M.

(iii) Preparation of indicators: As in sections 17.1 and 17.3.

(iv) Preparation of buffer solution (pH = 10): As in section 17.1.

(v) Preparation of 2M NaOH: 80 g of sodium hydroxide is dissolved in distilled water and the volume is made 1000 c.c.

(vi) Estimation of total hardness (temporary and permanent) of tap water can be done from the first titration of this experiment assuming the given solution to be tap water.

Hardness of tap water is determined in ppm assuming all the hardness to be present as $CaCO_3$. ppm may be defined as under:

Number of gm of solute present in 1000 litre of solution is known as ppm (parts per million).

In other words, number of mg of solute present in 1 litre of solution is known as ppm.

$$Na_2H_2Y + CaCO_3 \rightarrow Na_2CaY + H_2CO_3$$

372 g $Na_2H_2Y \equiv$ 100 g $CaCO_3$

or, $\frac{372}{100} g \equiv \frac{100}{100} g$

or, 3.72 g $Na_2H_2Y \equiv$ 1 g $CaCO_3$

$\because$ 1000 c.c. of 0.01M EDTA $\equiv$ 1 g of $CaCO_3$

$\therefore$ 1 c.c. of 0.01M EDTA $\equiv$ 0.001 g of $CaCO_3$

$\equiv$ 1 mg of $CaCO_3$

$\therefore$ 25 c.c. of 0.01 M EDTA

$\equiv (1 \times 25)$ mg of $CaCO_3$

$\equiv$ 25 mg of $CaCO_3$

$\because$ 25 c.c. tap water contains 25 mg of $CaCO_3$

$\therefore$ 1000 c.c. tap water contains $\frac{25 \times 1000}{25}$

= 1000 mg $CaCO_3$

Therefore, total hardness = 1000 ppm

To determine the permanent hardness of water, it is boiled vigorously for 30 minutes, cooled and filtered. Now taking 25 c.c. (or 50 c.c.) of this water and titrating it with 0.01M EDTA and EBT/KNO_3 as indicator, permanent hardness can be calculated. And thus by subtracting permanent hardness from total hardness, temporary hardness can also be determined.

17.5. To determine Aluminium in a given solution using 0.01M EDTA by back titration

Theory: First of all excess of 0.01M EDTA solution (definite volume) is added to the aluminium ion solution. Most of the EDTA forms complexes with aluminium ions.

$$H_2Y^{2-} + Al^{3+} \rightarrow AlY^- + 2H^+$$

$$372 \text{ g } Na_2H_2Y \equiv 27 \text{ g Al}$$

or, $\frac{372}{100}\text{g} = \frac{27}{100}\text{g}$

or, $3.72 \text{ g } Na_2H_2Y = 0.27 \text{ g Al}$

or, 1000 c.c. of 0.01M EDTA $\equiv$ 0.27 g of Al

or, 1 c.c. of 0.01M EDTA $= \frac{0.27}{1000}$ g of Al

$$= \frac{0.27 \times 1000}{1000} \text{ mg of Al}$$

or, 1 c.c. of 0.01M EDTA $\equiv$ 0.27 mg of Al

Remaining EDTA is titrated with 0.01M zinc sulphate solution. Difference between volumes of 0.01M EDTA added and 0.01M zinc sulphate used gives the volume of 0.01M EDTA consumed to form complex and 1 c.c. of consumed 0.01M EDTA will be equivalent to 0.27 mg of aluminium.

1 c.c. of consumed 0.01M EDTA $\equiv$ 0.27 mg of Al

Procedure: First of all 25 c.c. of aluminium ion solution (approximately 0.01M) is pipetted out and taken into a conical flask. 30 c.c. of 0.01M EDTA is added to it by a burette. The pH of the solution is adjusted between 7 and 8 by the addition of ammonia solution. The solution is boiled for a few minutes to ensure complete complexation of aluminium. It is cooled to room temperature and pH is adjusted between 7 and 8. About 50 mg EBT/KNO_3 mixture is added and titrated rapidly with 0.01M zinc sulphate solution until the colour changes from blue to wine red.

Observation and Calculation:

Table 44

Vol. of aluminium ion solution taken = 25 c.c.

0.01 M EDTA solution added = 30 c.c.

Sl. No.	Vol. of solution to be titrated	Vol. of 0.01M zinc sulphate taken in burette (c.c.)			Concordant reading (c.c.) (say)	Indicator	End point
		Initial reading (say)	Final reading (say)	Diff. (say)			
1.	Remaining EDTA	0.0	5.1	5.1			
2.	Remaining EDTA	5.1	10.1	5.0	5.0	EBT	Wine-
3.	Remaining EDTA	10.1	15.1	5.0			red

Vol. of aluminium ion solution = 25 c.c.

Vol. of 0.01M EDTA added = 30 c.c.

Vol. of 0.01M EDTA consumed to form complex with Al^{3+}

= Vol. of 0.01M EDTA added

− Vol. of 0.01M zinc sulphate used

= (30 − 5) c.c.

= 25 c.c.

$\because$ 1 c.c. of 0.01M EDTA ≡ 0.27 mg of Al

$\therefore$ 25 c.c. of 0.01M EDTA ≡ (0.27 × 25) mg of Al

= 6.75 mg of Al ... (1)

From titration,

25 c.c. of aluminium ion solution

≡ 25 c.c. of 0.01M EDTA ... (2)

From equations (1) and (2),

25 c.c. of aluminium ion solution ≡ 6.75 mg of Al

$\therefore$ 1000 c.c. of aluminium ion solution

$$= \frac{6.75 \times 1000}{25} \text{ mg of Al}$$

$$= \frac{6.75 \times 1000}{25 \times 1000} \text{ g of Al}$$

Therefore, amount of Al = 0.27 g/litre

Things to Remember

(i) Aluminium ion solution should be neutral.

(ii) Remaining EDTA may be titrated with magnesium sulphate or zinc sulphate using EBT as an indicator.

5 GRAVIMETRIC ANALYSIS

1. INTRODUCTION

Gravimetric analysis is a process in which the amount of a substance present is determined by converting it, by a suitable chemical reaction, into some other substance of known chemical composition, which can be readily isolated, purified, and weighed. It is a type of quantitative analysis.

Gravimetric analysis gives accurate results but results are obtained after a variety of operations like preparation of solutions, precipitation, filtration, ignition or drying, etc. That is why the gravimetric analysis becomes laborious (more time consuming) and highly technical. Gravimetric analysis involves the following operations:

1.1. Weighing of the Sample

A clean dry stoppered weighing bottle is partially filled with the given sample and weighed keeping it in a pan. The tube is removed from the pan, unstoppered and the desired amount of the substance is poured into a beaker. It is further weighed. The difference of the two weights give the weight of the sample taken in beaker. If the substance be not hygroscopic, it may be weighed taking on watch glass (clock glass).

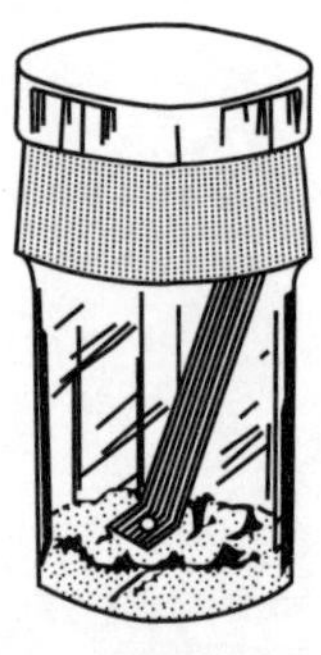

Fig. 1. Stoppered weighing tube

1.2. Preparation of Solution

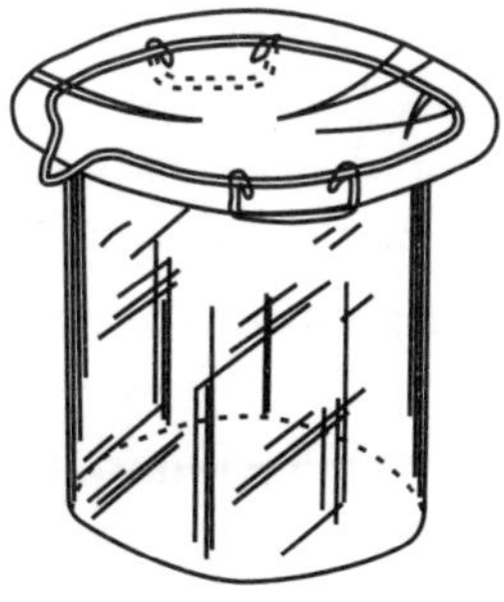

Fig. 2. Solution in a 400 ml. beaker covered with a clock glass

The weighed substance is dissolved in distilled water, sometimes dilute hydrochloric, nitric or acetic acid is also to be added. But, generally, the students are supplied with the prepared solutions. A definite amount of the solution must give a precipitate weighing between 0.2 g to 0.3 g.

1.3. Precipitation

The precipitate in gravimetric determinations should have the following characteristics:

(i) It must be highly insoluble so that the constituent being estimated is quantitatively precipitated.

(ii) The precipitate should be filterable.

(iii) It must have a known stable composition itself or it may be easily converted into a weighing form of definite known composition.

(iv) It should be of high purity, that is free from co-precipitated substances.

1.3.1. Co-precipitation

The process by which soluble impurities become incorporated into a precipitate particle during its formation is known as co-precipitation. This occurs due to adsorption or occlusion.

1.3.2. Post-precipitation

This is the special type of precipitation which occurs on the surface of the first precipitate after its formation.

1.3.3. Optimum conditions for precipitation

Following conditions tend to yield large particles and minimum contamination of precipitate:

(i) **Dilute solution of reagents:** Precipitation should be carried out from dilute solution of reagents. This removes the error due to co-precipitation.

(ii) **Slow addition of reagent:** A dilute solution of precipitating reagent should be added slowly with constant stirring of the solution.

(iii) **Precipitation from hot solution:** Precipitation is usually made in hot solution. This checks post-precipitation and helps in coagulation and better growth of crystals of the precipitate.

(iv) **An excess of the reagent:** A slight excess of the reagent is required to ensure complete precipitation. However, a large excess is to be avoided as it may lead to increasing solubility or contamination of the precipitate.

(v) **Reprecipitation:** It is sometimes necessary to obviate post-precipitation. The precipitate first formed is filtered and dissolved in some acid and then precipitated again.

1.4. Digestion (Ageing)

The precipitate is usually allowed to stand either at room temperature or at a flame or on a water bath (depending upon the nature of the precipitate) for a few hours (12-24). This is called digestion or ageing of the precipitate and this effects complete precipitation which occurs bit slowly and it also helps in obtaining a granular dense precipitate, in a readily filterable form. But unfortunately no one method can be applied for the digestion of all the precipitates. Therefore, ageing on a steam bath or on a hot plate quite generally results in the purification of the original precipitate. During the process of digestion of precipitates, at least two changes occur. The smaller precipitate particles which have a greater solubility than the larger ones, tend to pass into solution, and will ultimately redeposit upon the larger particles; therefore co-precipitation on smaller particles is thus eliminated and thus

Fig. 3. Digestion of the precipitate on steam-bath

the total co-precipitation on the ultimate precipitate is reduced. The net result of digestion is usually to reduce the extent of co-precipitation and to increase the size of the particles, rendering filtration easier. Digestion has little effect upon amorphous or gelatinous precipitates.

1.5. Filtration

Filtration involves the isolation of the precipitate from the mother liquor. Filtration is the separation of a solid from a liquid by means of a porous material. Following four methods of filtration have been described:

1.5.1. Filter funnel

The simplest apparatus used for filtration is the filter funnel fitted with a Whatman filter paper. The funnel should have an angle as close to 60° as possible, and a long stem (15 cm) to promote rapid filtration. A suitable filter paper is selected, folded exactly in halves and then in quarters so that the two halves do not exactly coincide. It is then opened with three quarters on one side and single quarter on the other and then placed in the funnel. The filter paper is wetted with distilled water and well pressed to fit tightly against the wall of the funnel. This funnel is then hanged in vertical position placing an empty beaker below it such that the funnel stem is always in touch with its wall and above the bottom.

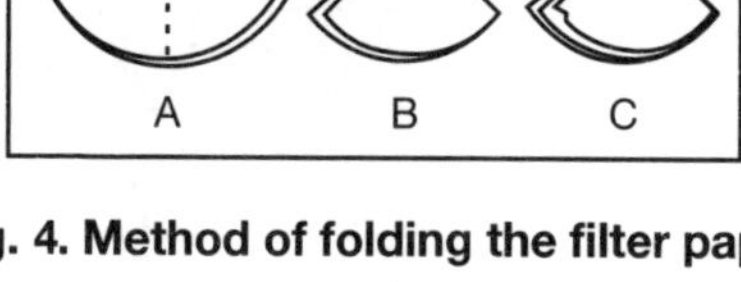

Fig. 4. Method of folding the filter paper

The supernatant liquid is allowed to pour down along the stirring glass rod against the side of the filter funnel. The liquid falling against the side and not the apex of the filter paper, should not fill in completely but a half cm below the top of the filter paper. If filter paper is fitted tightly in the funnel, there will be a continuous flow of liquid in the stem of funnel which accelerates a quick filtration.

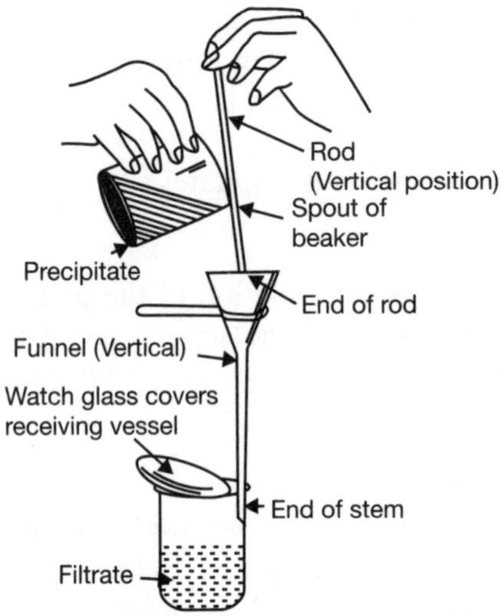

Fig. 5. Method of transferring the precipitate to the filter paper

Sometimes a little of precipitate or some of its particles remain sticking with the side of the beaker. This can be removed by rubbing around the inner surface of the beaker with a rubber tipped glass rod known as 'policeman'.

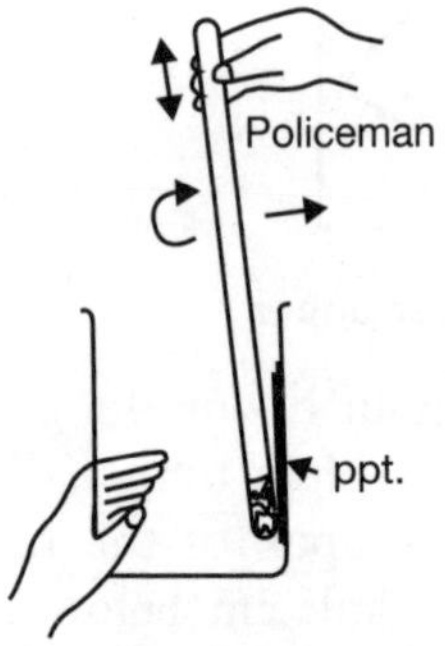

Fig. 6. How to use a policeman

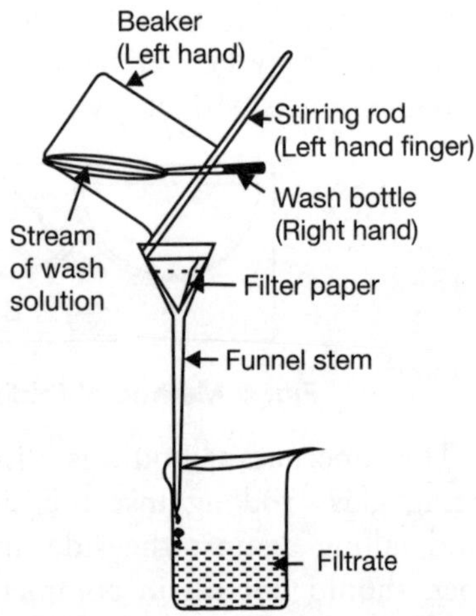

Fig. 7. Method of transferring the precipitate to the filter paper

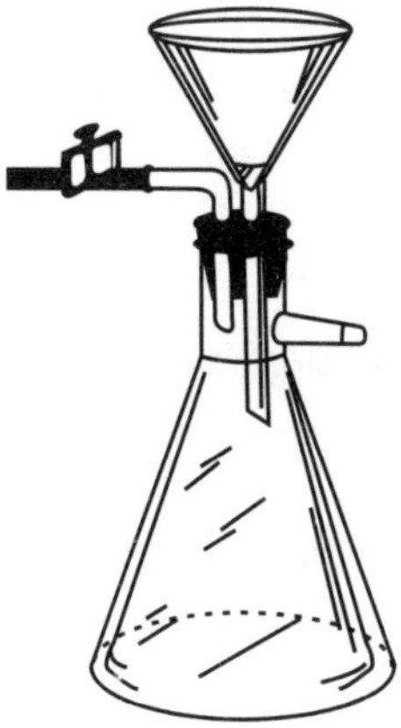

Fig. 8. Arrangement for suction filtration with filter paper

1.5.2. Gooch crucible

F.A. Gooch (1878) first employed a filter mat of purified asbestos, this was supported inside a tall form of platinum crucible, the bottom of which was perforated with numerous holes. This kind of crucible has enjoyed great popularity. The filtering medium is a mat of asbestos fibres made by pouring suspension of asbestos into the crucible. On applying suction the fibres are drawn to the bottom and form a compact mat which will retain even fine grained precipitates.

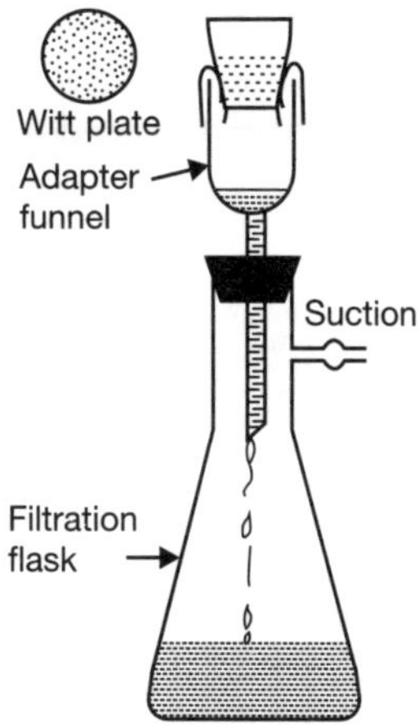

Fig. 9. Arrangement for suction filtration with Gooch and sintered glass crucibles

A clean Gooch crucible is placed in an adapter funnel attached to a suction flask and is fitted with a thin suspension of asbestos. Slow suction is applied, and more asbestos poured in if necessary, so that a mat 1 to 2 mm in thickness is finally formed. A perforated porcelain plate (witt plate) is now placed upon the asbestos mat. The witt plate is used to protect the asbestos mat when liquid is poured into the crucible. The suction is increased, and water is passed through the crucible until fine fibres are no longer visible in the filtrate when examined in good light. The crucible is then dried to constant weight at the same temperature to which the precipitate will be heated later.

1.5.3. Sintered glass crucible

Sintered glass crucible possesses an advantage over Gooch crucible as it does not require the preparation of a filter mat. It is made of resistance glass and have a porous disc of sintered ground glass fused into the body of the crucible. The filter disc is made in varying porosities as indicated by number from 0 (the coarsest) to 5 (the finest); the range of pore diameter for the various grades is as follows:

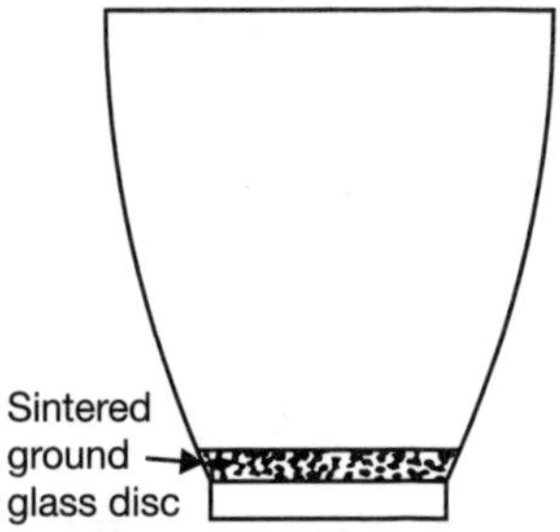

Fig. 10. Sintered glass crucible

Porosity	Pore diameter (μm)
0	200 – 250
1	100 – 120
2	40 – 50
3	20 – 30
4	5 – 10
5	1 – 2

Porosity 3 is suitable for precipitates of moderate particle size, and porosity 4 for fine precipitates such as barium sulphate. This crucible should not be heated above about 200°C.

Silica crucible of similar pattern are also available, and, although expensive, has certain advantage in thermal stability.

While using sintered glass crucible for filtration not more than three fourth (3/4th) of its volume should be filled with the liquid.

1.5.4. Buchner funnel

For filtering large quantities of material, a Buchner funnel is usually employed. It requires Whatman filter paper for filtration.

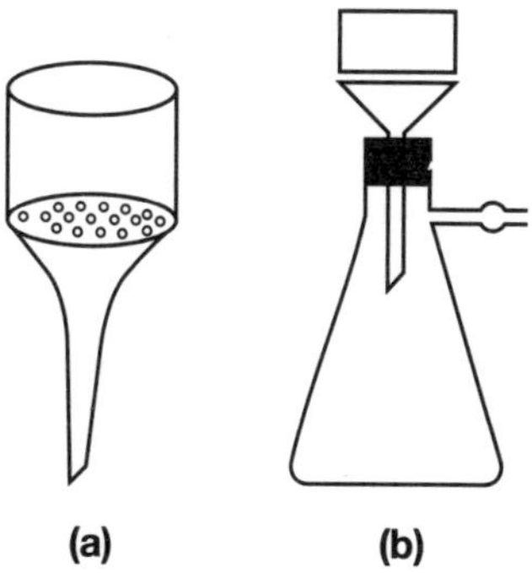

Fig. 11

1.6. Washing

The object of washing is to free the precipitate from the dissolved substances or adsorbed ions with the minimum amount of the washing liquid and to reduce the loss of precipitate by solution. The washing liquid should have a minimum solvent effect on the precipitate and should not help in peptizing it even after electrolytic ions have been washed off. Different kind of wash solutions to be used depends upon the precipitate (and to a lesser extent upon the substances in solution). The factors that must be considered especially are the solubility of the precipitate and its peptizability. If a precipitate is appreciably soluble in water, it cannot, of course, be washed with this liquid. This difficulty can be overcome, when a cation is being

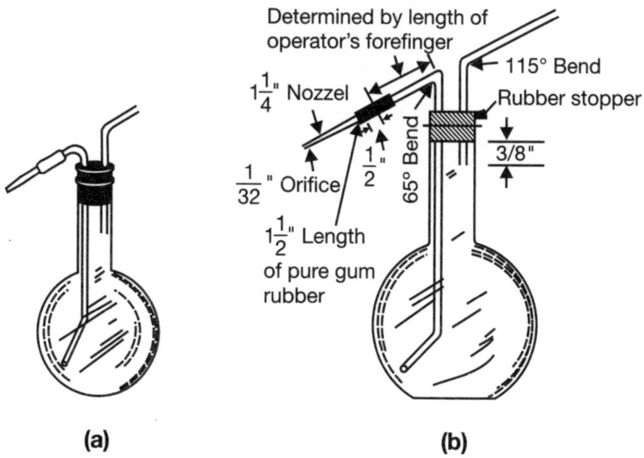

Fig. 12. Wash bottles

determined, by washing the precipitate with a dilute solution of an ammonium salt having an ion in common with precipitate. The ammonium salt remaining with the precipitate after washing is volatilized on ignition at a low temperature.

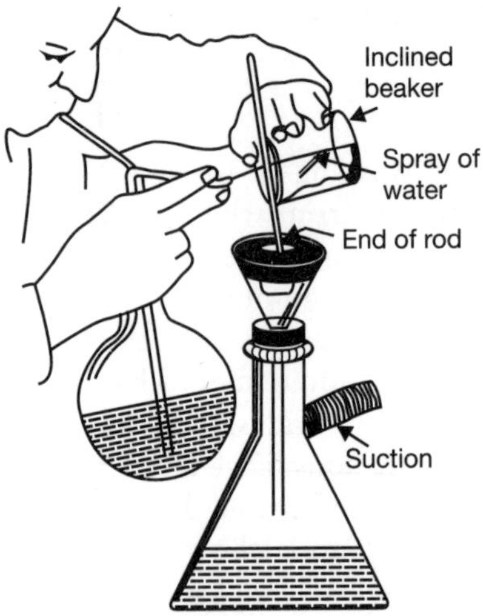

Fig. 13. Method of washing

1.6.1. Qualities of an ideal washing liquid

An ideal washing liquid must possess the following qualities:

(i) It should have no solvent action upon the precipitate, but dissolve readily foreign substances.

(ii) It should have no peptizing action upon the precipitate.

(iii) It should not form any chemical compound with the precipitate.

(iv) It should be easily volatile at the temperature of drying of the precipitate.

(v) It should not contain any such substance which may interfere with subsequent determinations in the filtrate.

1.6.2. Classification of wash solutions

Wash solutions can be conveniently divided in three classes:

(i) First group of wash solutions are the solutions which prevent the precipitate from becoming colloidal and passing through the filter. Dilute ammonium nitrate solution is employed for washing ferric hydroxide and 1% nitric acid for washing silver chloride.

(ii) Second group of wash solutions are the solutions which reduce the solubility of the precipitate. The wash solutions containing a common ion with the precipitate may also appreciably reduce the solubility of the precipitate.

(iii) Third group of wash solutions are the solutions which prevent the hydrolysis of salts of weak acids and bases. The addition of an acid to the wash solution prevents the hydrolysis of ferric or similar salts.

1.7. Drying

The filter is allowed to drain completely and the water adhering in the stem of the funnel is removed with a piece of filter paper. The funnel with the filter paper is covered by a piece of paper with holes in it. It is then placed over a truncated iron cone and heated over an asbestos wire gauze. When the filter paper gets loosened, the drying is considered to be complete. Overheating is always avoided as it chars the filter paper.

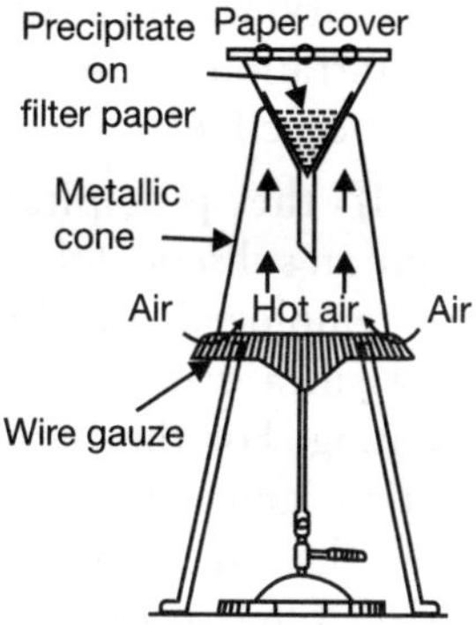

Fig. 14. Drying of the ppt. on an iron cone

Sintered glass crucible or funnel containing precipitate is allowed to stand in an electric oven (temperature of which is maintained) for drying.

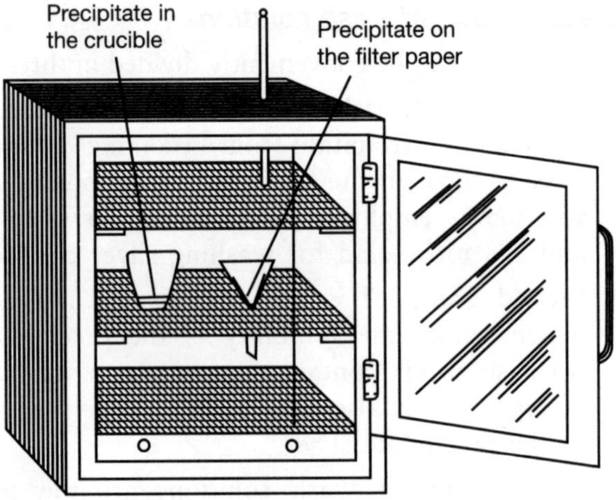

Fig. 15. Electric oven/Air oven

1.8. Ignition (or, Incineration)

If the precipitate does not require ignition and it requires only drying, then ideal crucible for filtration is Gooch or sintered glass crucible.

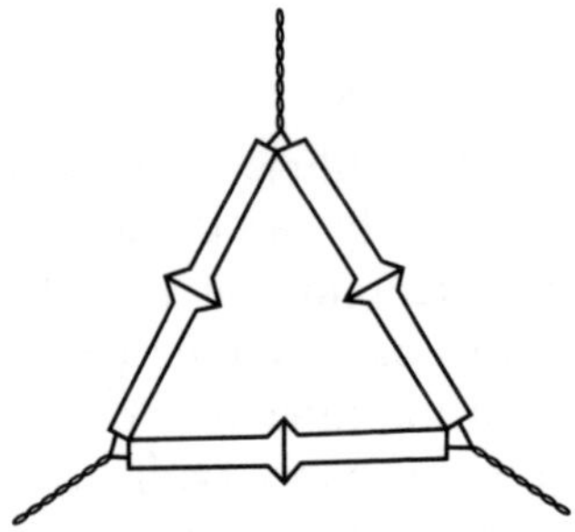

Fig. 16. Clay pipe triangle

If the precipitate requires ignition subsequently, filter papers are employed as filtering media and silica crucibles are used for igniting. For this a silica crucible is first ignited to constant weight (i.e., heating, cooling and weighing). The dried filter paper containing precipitate is folded so as to completely enclose the precipitate, care being taken not to tear the paper. The packet is then placed in the weighed crucible which is supported on a clay pipe triangle as shown in the figure. The heating is started slowly with a small flame. The filter paper is burnt off. The paper should smoulder off and never be permitted to catch on the

Fig. 17. Crucible and lid

fire. After the paper has been burnt off, the crucible is placed in the upright position. Now the heating is done at the desired temperature and is continued till required.

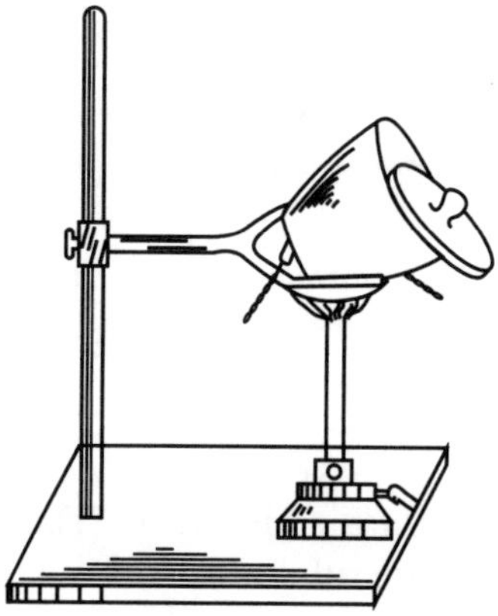

Fig. 18. Method of ignition

1.9. Weighing of the Product

The hot product is cooled in desiccator and weighed. Further the product is heated, cooled and weighed. The process is repeated till constant weight is obtained.

Fig. 19. Desiccator

2. EXPERIMENTAL

2.1. To estimate gravimetrically the amount of barium per litre in the supplied barium chloride solution

Theory: Barium is precipitated as barium sulphate by the addition of hot dilute sulphuric acid to the hot and dilute solution of barium chloride. The white precipitate is filtered, washed till free from sulphate and chloride ions, dried, ignited and weighed as $BaSO_4$.

$$BaCl_2 + H_2SO_4 \rightarrow \underset{137+32+64}{BaSO_4} \downarrow + 2HCl$$

$$= 233$$

$\because$ 233 g of $BaSO_4$ contains 137 g of barium

$\therefore$ 1 g of $BaSO_4$ contains $\frac{137}{233}$ g of barium

= 0.588 g of barium

Procedure: At first 25 c.c. of the given barium chloride solution is taken in a beaker. 0.5 c.c. of conc. HCl and 100 c.c. of distilled water are added to it. The solution is heated to boiling and hot dilute sulphuric acid is added dropwise with constant stirring till the precipitation is complete. When the precipitation appears to be complete, it is allowed to settle down and half a c.c. more sulphuric acid is added. If no turbidity appears even after waiting for a minute, the precipitation may be taken to be complete. To ensure it, extra 5 c.c. of sulphuric acid is added with vigorous stirring and digested on water bath for about half an hour. Now it is filtered with the help of Whatman filter paper No. 42. The residue is washed with hot distilled water till the filtrate becomes free from Cl^- and SO_4^{2-} ions. Then the residue with funnel is kept in air oven for drying. After drying, the filter paper is removed from the funnel. The precipitate is collected in a silica crucible placed on a glazed paper with the help of a fine feather. The filter paper is folded several times to make it look like a cone. Now it is held by means of a pair of tongs above the crucible and burnt with a flame of the burner keeping it horizontally. Ash is collected in the crucible. Now the crucible is ignited first gently and then strongly. At last a few drops of conc. H_2SO_4 are added and ignited further. The crucible is cooled first in air (2-3 minutes) and then in desiccator. It is weighed. Heating, cooling and weighing are continued till constant weight is obtained. The silica crucible alone is weighed before using in the ignition.

Calculation:

Weight of silica crucible = 30.215 g (say)

Weight of silica crucible + $BaSO_4$ = 30.502 g (say)

$\therefore$ Wt. of $BaSO_4$ = 0.287 g

$\because$ 233 g barium sulphate contains 137 g of barium

$\therefore$ 0.287 g barium sulphate contains $\frac{137 \times 0.287}{233}$ g of barium = 0.169 g of barium

$\because$ 25 c.c. of supplied solution contains 0.169 g of barium

$\therefore$ 1000 c.c. of supplied solution contains $\frac{0.169 \times 1000}{25}$ g of barium = 6.76 g of barium

Hence the amount of barium = 6.76 g/litre.

Things to Remember

(i) Preparation of barium chloride solution (for teachers): 12.5 g of barium chloride ($BaCl_2 \cdot 2H_2O$) is dissolved in distilled water and the volume is made 1000 c.c.

(ii) Precipitating reagent (N – H_2SO_4): 1 c.c. of concentrated sulphuric acid is diluted to 30 c.c. with distilled water.

(iii) Wash solution: Hot distilled water is used as wash solution.

(iv) Whatman filter paper is ashless.

(v) During ignition $BaSO_4$ is slightly reduced to BaS by the carbon of the filter paper.

$$BaSO_4 + 4C \rightarrow BaS + 4CO\uparrow$$

So, at least a few drops of conc. H_2SO_4 are added with the help of a jet on the precipitate in the crucible in cold condition and again heated.

$$BaS + H_2SO_4 \rightarrow BaSO_4 + H_2S\uparrow$$

(vi) The precipitation should be carried out in presence of about 0.05N HCl solution which helps to increase the size of the precipitate particles. This makes the precipitate easily filterable.

(vii) Large excess of HCl should not be added as $BaSO_4$ dissolves in it owing to formation of the bisulphate.

$$2BaSO_4 + 2H^+ \rightarrow \underset{\text{(Soluble)}}{Ba(HSO_4)_2} + Ba^{2+}$$

(viii) In order to prevent the co-precipitation of $BaCl_2$ with $BaSO_4$ the precipitation must be performed by adding hot precipitant to hot solution, very slowly with vigorous stirring.

(ix) Barium nitrate cannot be used in place of the chloride as it does not prevent co-precipitation with barium sulphate. For the same reason, the nitrate ions, if present, must be removed by heating it with excess of HCl before the precipitation of $BaSO_4$.

(x) The precipitate of $BaSO_4$ may be filtered by prior weighed sintered glass crucible, G-4 using gentle suction. It is dried in air over for 1 hour at 100-110°C. This method is easier than the method applied in the present experiment.

2.2. To estimate sulphate ion gravimetrically in the given solution of copper sulphate

Theory: Sulphate ion is precipitated as barium sulphate by the addition of hot and dilute solution of barium chloride to the hot solution of any salt containing sulphate ions.

$$CuSO_4 + BaCl_2 \rightarrow \underset{137+32+64}{BaSO_4} \downarrow + CuCl_2$$

$$= 137 + 96$$

$$= 233$$

$\because$ 233 g of $BaSO_4$ contains 96 g of sulphate ion

$\therefore$ 1 g of $BaSO_4$ contains $\frac{96}{233}$ g of sulphate ion

$= 0.412$ g of sulphate ion

Procedure: 25 c.c. of the supplied solution of copper sulphate is taken in a beaker. It is diluted to 150 c.c. with distilled water. 2 c.c. of dilute hydrochloric acid is added to it. The solution is heated to boiling and 5% barium chloride solution is added dropwise with constant stirring till the precipitation is complete. The resulting solution is heated on a very small flame or on a water bath for an hour. The precipitate is allowed to settle. The supernatant liquid is tested by adding a few drops of barium chloride solution for complete precipitation. The solution is filtered with the help of Whatman filter paper No. 42. The residue is washed 2-3 times with hot distilled water. It is dried in air oven and the filter paper with precipitate is ignited taking it in pre weighed silica crucible. Finally a few drops of conc. H_2SO_4 are added, heated, dried and weighed.

Calculation:

Wt. of silica crucible = 30.215 g (say)

Wt. of silica crucible + $BaSO_4$ = 30.505 g (say)

$\therefore$ Wt. of $BaSO_4$ = 30.505 – 30.215

= 0.290

$\because$ 1 g of $BaSO_4$ contains 0.412 g of SO_4^{2-} ion.

$\therefore$ 0.29 g of $BaSO_4$ contains 0.412 × 0.29 g of SO_4^{2-} ion.

= 0.119 g of SO_4^{2-} ion.

$\because$ 25 c.c. of supplied solution contains 0.119 g of SO_4^{2-} ion.

$\therefore$ 1000 c.c. of supplied solution contains $\frac{0.119}{25} \times 1000\,\text{g}$ of SO_4^{2-} ion = 4.76 g of SO_4^{2-} ion

Therefore, amount of sulphate ion = 4.76 g/litre.

Things to Remember

(i) Preparation of copper sulphate solution (for teachers): 14 g of copper sulphate, $CuSO_4 \cdot 5H_2O$ is dissolved in distilled water and the volume is made 1000 c.c.

(ii) Preparation of barium chloride solution: 5 g of barium chloride is dissolved in 100 c.c. of distilled water to get 5% solution.

(iii) Wash solution: Hot distilled water.

(iv) Barium nitrate cannot be used instead of barium chloride.

(v) Sodium sulphate solution, dil. H_2SO_4 or any soluble sulphate may be taken instead of copper sulphate.

(vi) The precipitate may be filtered by sintered glass crucible.

2.3. To estimate silver gravimetrically in the given solution of silver nitrate

Theory: Silver is precipitated as silver chloride by the addition of dil. hydrochloric acid to the warm silver nitrate solution.

$$AgNO_3 + HCl \rightarrow \underset{108+35.5}{AgCl} \downarrow + HNO_3$$

$$= 143.5$$

$\because$ 143.5 g of AgCl contains 108 g of silver

$\therefore$ 1 g of AgCl contains $\frac{108}{143.5}$ g of silver

$= 0.753$ g of silver

Procedure: 25 c.c. of the given silver nitrate solution is taken in a beaker. It is diluted to 150 c.c. and 0.5 c.c. of concentrated HNO_3 is added to it. The solution is heated to 70°C. Now 0.2N HCl is added slowly with constant stirring till no further precipitation takes place (5-10 c.c. of 0.2N HCl is sufficient for complete precipitation). The precipitate is warmed for 5 minutes and allowed to cool at room temperature in a dark place. Now the supernatant liquid is tested by adding a few drops of 0.2N HCl along the stirring glass rod to ensure complete precipitation. The precipitate is allowed to settle down in dark place for at least 2 hours or preferably overnight.

Now it is filtered through pre weighed sintered glass crucible G-3. First supernatant liquid is poured. The precipitate is washed (3-4 times) by decantation with 0.1N HNO_3. Then the precipitate is transferred to the crucible with the help of policeman. Now the precipitate is washed with 0.01N HNO_3 till it becomes free from Cl^- ion. Finally, the precipitate is washed with distilled water to remove HNO_3. It is dried in air oven, cooled and weighed. The process is repeated till constant weight is obtained.

Calculation:

Wt. of sintered glass crucible = 24.576 g (say)

Wt. of sintered glass crucible + AgCl = 24.955 g (say)

$\therefore$ Wt. of AgCl = 0.379 g

$\because$ 1 g of AgCl contains 0.753 g of silver

$\therefore$ 0.379 g of AgCl contains 0.753 × 0.379 g of silver

= 0.285 g of silver

$\because$ 25 c.c. of $AgNO_3$ solution contains 0.285 g of silver

$\therefore$ 1000 c.c. of $AgNO_3$ solution contains $\frac{0.285}{25} \times 1000\,g$ of silver = 11.4 g of silver

Therefore, amount of silver = 11.4 g/litre.

Things to Remember

(i) Preparation of silver nitrate solution (for teachers): 18 g of silver nitrate is dissolved in distilled water in presence of a little nitric acid and the volume is made 1000 c.c.

(ii) Precipitation reagent (0.2N HCl): 2 c.c. of conc. HCl is diluted to 100 c.c. with distilled water.

(iii) Wash solutions:

- 0.1N HNO_3: 3 c.c. of conc. HNO_3 is diluted to 500 c.c. with distilled water.
- 0.01N HNO_3: 10 c.c. of 0.1N HNO_3 is diluted to 100 c.c. with distilled water.

(iv) The precipitation should be carried out in presence of HNO_3. It helps in coagulation of the precipitate and prevents the co-precipitation of CO_3^{2-}, PO_4^{3-}, etc. when present as impurities. Moreover, it reduces the solubility of silver chloride.

(v) During precipitation, dil. HCl should be added slowly. Large excess of it should not be added as the precipitated silver chloride may dissolve forming a complex.

$$AgCl + \underset{(excess)}{Cl^-} \rightarrow \underset{[\text{dichloro argentate (I) ion}]}{[AgCl_2]^-}$$

(vi) The precipitated silver chloride must be protected from light as silver chloride decomposes into silver and chlorine in presence of light.

$$2AgCl \xrightarrow{\text{light}} 2Ag + Cl_2$$

(vii) The precipitate should be dried below 120°C as it can be volatilized or decomposed at higher temperature.

(viii) The precipitate should be washed with water containing nitric acid to prevent peptization.

(ix) Generally tap water contains appreciable amount of chloride. Hence use of distilled water is essential.

(x) Since silver chloride is easily reduced into metallic silver, hence the precipitate should be filtered through sintered glass crucible.

2.4. To estimate chloride ion gravimetrically in the given solution of sodium chloride

Theory: Chloride is precipitated as silver chloride by the addition of 0.1N $AgNO_3$ solution.

$$NaCl + AgNO_3 \rightarrow \underset{108+35.5}{AgCl} \downarrow + NaNO_3$$

$$= 143.5$$

$\because$ 143.5 g of silver chloride contains 35.5 g of Cl^- ion.

$\therefore$ 1 g of silver chloride contains $\frac{35.5}{143.5}$ g of Cl^- ion

$= 0.247$ g of Cl^- ion.

Procedure: First of all 25 c.c. of supplied solution of sodium chloride is taken in a beaker and diluted to 50 c.c. with distilled water. 0.5 c.c. of conc. HNO_3 is added to it. The solution is heated to 70°C. Now 0.1N $AgNO_3$ solution is added slowly with constant stirring till no further precipitation occurs (5 - 10 c.c. of 0.1N $AgNO_3$ are sufficient for complete precipitation). It is warmed for five minutes and allowed to cool at room temperature in dark. Now the supernatant liquid is tested by adding few drops of 0.1N $AgNO_3$ solution to ensure complete precipitation. The precipitate is allowed to settle in a dark place for 2 hours or preferably overnight.

Now the supernatant liquid is poured through a previously weighed sintered glass crucible. The precipitate is washed by decantation with 0.01N HNO_3. Then the precipitate is transferred into the crucible and again washed with 0.01N HNO_3 until free from Ag^+ ions. Finally it is washed with distilled water to remove HNO_3. Now the crucible with the precipitate is dried at 120°C in air oven, cooled and weighed till constant weight.

Calculation:

Wt. of sintered glass crucible = 24.576 g (say)

Wt. of sintered glass crucible + AgCl = 24.786 g (say)

$\therefore$ Wt. of AgCl = 0.210 g

$\because$ 1 g of AgCl contains 0.247 g of Cl^- ion

$\therefore$ 0.210 g of AgCl contains 0.247 × 0.210 g of Cl^- ion

= 0.052 g of Cl^- ion

$\because$ 25 c.c. of the given solution contains 0.052 g of Cl^- ion

$\therefore$ 1000 c.c. of the given solution contains $\frac{0.052}{25} \times 1000$ g of Cl^- ion = 2.08 g of Cl^- ion.

Therefore, amount of chloride ion = 2.08 g/litre.

Things to Remember

(i) Preparation of sodium chloride solution (for teachers): 5 g of sodium chloride (A.R.) is dissolved in distilled water and the volume is made 1000 c.c.

(ii) Precipitation reagent (0.1N $AgNO_3$): 1.688 g of silver nitrate is dissolved in distilled water and the volume is made 100 c.c.

(iii) Wash solution (0.01N HNO_3): See section 2.3.

2.5. To estimate the amount of iron gravimetrically in the supplied solution of ferrous ammonium sulphate

Theory: Iron is estimated by precipitation as ferric hydroxide which is then ignited and weighed as Fe_2O_3 and from its weight amount of iron is calculated.

$$FeSO_4 \cdot (NH_4)_2 SO_4 \cdot 6H_2O \rightarrow FeSO_4 + (NH_4)_2 SO_4 + 6H_2O$$

$$FeSO_4 \rightarrow Fe^{2+} + SO_4^{2-}$$

$$Fe^{2+} \xrightarrow{\text{conc. } HNO_3} Fe^{3+} + e$$

$$Fe^{3+} + 3OH^- \rightarrow Fe(OH)_3 \downarrow$$

$$2Fe(OH)_3 \xrightarrow{\text{Ignition}} \underset{55.85 \times 2 + 16 \times 3}{Fe_2O_3} + \underset{\text{(Vap.)}}{3H_2O \uparrow}$$

$$= 111.70 + 48$$

$$= 159.7$$

$\because$ 159.7 g of Fe_2O_3 contains 111.7 g of iron

$\therefore$ 1 g of Fe_2O_3 contains $\dfrac{111.7}{159.7}$ g of iron

= 0.699 g of iron

Procedure: First of all 25 c.c. of the supplied solution (Mohr's salt solution) is taken in a beaker. 25 c.c. of distilled water and 10 c.c. of HCl (1:1) are added to make the solution clear. Now about 3 c.c. of conc. HNO_3 is added and boiled for conversion of Fe^{2+} ion into Fe^{3+} ion until the solution becomes yellow. It is diluted to 200 c.c. with distilled water. The solution is boiled and then ammonia solution (1:1) added till alkaline. It is boiled for 1 minute and the precipitate allowed to settle. Now it is filtered through Whatman filter paper No. 40. The residue is washed with hot 1% ammonium nitrate solution 4 or 5 times.

Removal of chloride and sulphate ions are tested by adding $AgNO_3$ and $BaCl_2$ solutions respectively in the washings. Now the residue is dried in the air oven. After drying it is taken in a previously weighed silica crucible and ignited when reddish brown mass of Fe_2O_3 is obtained. Finally one drop of conc. HNO_3 is added and ignited for 15 minutes. It is cooled and weighed.

Calculation:

Wt. of crucible = 30.217 g (say)

Wt. of crucible + Fe_2O_3 = 30.462 g (say)

$\therefore$ Wt. of Fe_2O_3 = 0.245 g

$\because$ 1 g of Fe_2O_3 contains 0.699 g of iron

$\therefore$ 0.245 g of Fe_2O_3 contains 0.699 × 0.245 g of iron
= 0.171 g of iron

$\because$ 25 c.c. of the given solution contains 0.171 g of iron

$\therefore$ 1000 c.c. of the given solution contains $\frac{0.171}{25} \times 1000$ g of iron = 6.84 g of iron

Therefore, amount of iron = 6.84 g/litre.

Things to Remember

(i) If iron is to be estimated in ferrous salt, it is necessary to oxidise the latter with conc. HNO_3 before the precipitation.

(ii) The precipitate must be obtained in coagulated form, otherwise, it will be difficult to filter. This is done by affecting the precipitation at or near the boiling point.

(iii) Ammonium nitrate solution should be used as wash solution so as to prevent the peptization and the formation of slimy material.

(iv) Preparation of Mohr's salt solution (for teachers): 50 g of ferrous ammonium sulphate, $FeSO_4 \cdot (NH_4)_2SO_4 \cdot 6H_2O$ is dissolved in distilled water and the volume is made 1000 c.c. by adding water and a little dil. H_2SO_4.

(v) Preparation of precipitation reagent (1:1 ammonia solution): 50 c.c. of ammonia solution is added to 50 c.c. of distilled water.

(vi) Preparation of wash solution (1% NH_4NO_3 solution): 1 g of ammonium nitrate solution is dissolved in distilled water and the volume is made 100 c.c.

2.6. To estimate nickel gravimetrically in supplied solution of nickel salt

Theory: Nickel is precipitated as bis dimethylglyoximato nickel(II) complex by the addition of an alcoholic solution of dimethylglyoxime to a neutral or slightly acidic hot solution of the nickel salt, and then adding a slight excess of NH_4OH. A scarlet red crystalline precipitate is obtained, which is weighed as bis dimethylglyoximato nickel(II) after drying it at 110–120°C.

$$Ni^{2+} + 2\begin{bmatrix} CH_3-C=N-OH \\ | \\ CH_3-C=N-OH \end{bmatrix} \xrightarrow[-2H^+]{+NH_4OH,}$$

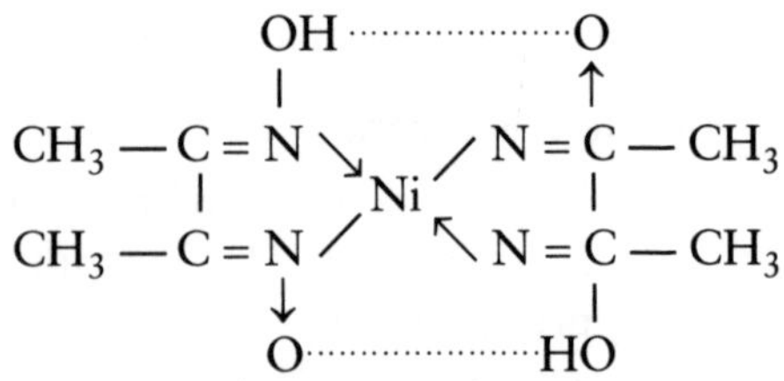

(Scarlet red ppt.)

$$\text{or, } Ni^{2+} + 2C_4H_8N_2O_2 \xrightarrow[-2H^+]{} \underset{96+14+56+64+58.7}{(C_4H_7N_2O_2)_2Ni}$$

$$= 288.7$$

∵ 288.7 g of nickel dimethylglyoximate contains 58.7 g of nickel.

∴ 1 g of nickel dimethylglyoximate contains $\frac{58.7}{288.7}$ g of nickel = 0.203 g of nickel

Procedure: First of all 25 c.c. of the supplied solution (under examination) is taken in a beaker. About 5 c.c. of dil. HCl (1:1) is added and diluted to 200 c.c. with distilled water. It is heated to 70–80°C. Now slight excess (30–40 c.c.) of dimethylglyoxime is added and then immediately dilute ammonia solution is added dropwise with constant stirring until precipitation takes place and the solution becomes alkaline. Then it is allowed to stand on steam bath for 20–30 minutes and tested for complete precipitation by adding a drop of dimethylglyoxime. Now the precipitate is allowed to stand for one hour and the cold solution filtered through previously weighed sintered glass crucible. The precipitate is washed with hot water till free from chloride ion. Now the precipitate is dried at 100–120°C for 40 to 45 minutes and weighed till constant weight after cooling.

Calculation:

Wt. of sintered glass crucible = 24.576 g (say)

Wt. of sintered glass crucible + nickel dimethylglyoximate = 24.926 g (say)

$\therefore$ Wt. of nickel dimethylglyoximate = 0.350

$\because$ 1 g of nickel dimethylglyoximate contains 0.203 g of nickel

$\therefore$ 0.350 g nickel dimethylglyoximate contains $0.203 \times 0.350 = 0.071$ g of nickel

$\because$ 25 c.c. of the given solution contains 0.071 g of Ni

$\therefore$ 1000 c.c. of the given solution contains $\frac{0.071}{25} \times 1000$ g of Ni = 2.84 g of Ni

Therefore, amount of nickel = 2.84 g/litre.

Things to Remember

(i) Preparation of nickel ammonium sulphate solution (for teachers): 18 g of nickel ammonium sulphate, $NiSO_4 \cdot (NH_4)_2SO_4 \cdot 6H_2O$ is dissolved in distilled water and a few c.c. of dil. hydrochloric acid (1:1). The volume is made 1000 c.c. with distilled water.

(ii) Preparation of precipitation reagent (1% alcoholic dimethylglyoxime solution): 1 g of dimethylglyoxime is dissolved in 100 c.c. of alcohol.

(iii) Preparation of ammonia solution (1:1): 50 c.c. of concentrated ammonia is added to 50 c.c. of distilled water.

(iv) Wash solution: Hot distilled water.

(v) Only a slight excess of dimethylglyoxime should be used, since it is not very soluble in water or in very dilute alcohol, and may precipitate, if large excess is added.

(vi) The precipitate should be dried at a temperature below 120°C as the precipitate tends to decompose at higher temperature.

2.7. To estimate copper gravimetrically in the given solution of copper salt, $CuSO_4 \cdot 5H_2O$

Theory: Copper is precipitated quantitatively as cuprous thiocyanate from slightly acidic solution in the presence of sulphurous acid using a solution of ammonium thiocyanate as reagent.

$$[Cu^{2+} + \cancel{e} \rightarrow Cu^{+}] \times 2$$

$$SO_3^{2-} + H_2O \rightarrow SO_4^{2-} + 2H^{+} + 2\cancel{e}$$

$$2\,Cu^{2+} + SO_3^{2-} + H_2O \rightarrow 2Cu^{+} + SO_4^{2-} + 2H^{+}$$

$$Cu^{+} + SCN^{-} \rightarrow \underset{\substack{\text{(White ppt.)} \\ 63.5 + 32 + 12 + 14}}{CuSCN \downarrow}$$

$$= 121.5$$

∵ 121.5 g of cuprous thiocyanate contains 63.5 g Cu

∴ 1 g of cuprous thiocyanate contains $\frac{63.5}{121.5}$ g of Cu

= 0.523 g Cu

Procedure: First of all 25 c.c. of the given copper salt solution is taken in a beaker. Few drops of dilute hydrochloric acid are added. Then slight excess (20–30 c.c.) of freshly prepared saturated sulphurous acid solution is added. Smell of SO_2 indicates the excess of sulphurous acid. The solution is diluted to about 150 c.c. It is heated to boiling. Now freshly prepared 10% ammonium thiocyanate solution (5–10 c.c.) is added slowly and with constant stirring to the hot solution. Large excess of the precipitant should be avoided. The precipitate (CuSCN) should be white and mother liquor should be colourless and with a smell of SO_2. In case the precipitate is coloured and the mother liquor is not colourless, more sulphurous acid solution should be added till the precipitate becomes white and then 2.5 c.c. of 10% ammonium thiocyanate solution should also be added. It is allowed to stand for 2-3 hours or preferably overnight. Now it is filtered through a previously weighed sintered glass crucible G-4 and the precipitate washed with ammonium thiocyanate and sulphurous acid solution (cold). Finally the precipitate is washed with 20% alcohol to remove ammonium thiocyanate. Then the precipitate is dried to constant weight at 105–120°C in air oven.

Calculation:

Wt. of sintered glass crucible = 24.576 g (say)

Wt. of sintered glass crucible + cuprous thiocyanate = 24.891 g (say)

∴ Wt. cuprous thiocyanate = 0.315 g

∵ 1 g of CuSCN contains 0.523 g of Cu

∴ 0.315 g of CuSCN contains 0.523 × 0.315 g of Cu = 0.165 g of Cu

∵ 25 c.c. of copper salt solution contains 0.165 g of Cu

∴ 1000 c.c. of copper salt solution contains $\frac{0.165}{25} \times 1000$ g of Cu = 6.6 g of Cu

Therefore, amount of copper = 6.6 g/litre.

Things to Remember

(i) Preparation of copper sulphate solution (for teachers): 26 g of copper sulphate, $CuSO_4 \cdot 5H_2O$ is dissolved in distilled water in presence of acetic acid and the volume is made 1000 c.c. by diluting it with distilled water.

(ii) Preparation of precipitation reagent (10% NH_4SCN): 10 g of ammonium thiocyanate is dissolved in distilled water and the volume is made 100 c.c.

(iii) Preparation of sulphurous acid solution: Sulphur dioxide gas is passed in 100 c.c. of distilled water till it is saturated with the gas.

(iv) Preparation of wash solutions:

- Ammonium thiocyanate and sulphurous acid solution: 100 c.c. of distilled water is taken in a wash bottle. 1 c.c. of 10% NH_4SCN solution and 5-6 drops of saturated sulphurous acid solution are added to it.
- 20% alcohol: 80 c.c. of distilled water is mixed with 20 c.c. of absolute alcohol.

(v) The copper salt solution must only be faintly acidic because the solubility of the precipitate increases appreciably with the concentration of hydrogen ions.

(vi) The wet precipitate of cuprous thiocyanate is slowly oxidised and for that reason a wash solution of 1% ammonium thiocyanate is used which contains a reducing agent like sulphurous acid.

2.8. To estimate zinc gravimetrically in the given solution

Theory: Zinc is precipitated as zinc ammonium phosphate, $ZnNH_4PO_4$ by adding diammonium hydrogen phosphate, $(NH_4)_2HPO_4$ to a slightly acidic solution of zinc salt. The precipitate is then weighed as $ZnNH_4PO_4$. The precipitant, $(NH_4)_2HPO_4$ exerts an important buffering action.

$$ZnCl_2 + (NH_4)_2 HPO_4 \rightarrow ZnNH_4PO_4 \downarrow + NH_4Cl$$
$$HCl + (NH_4)_2 HPO_4 \rightarrow NH_4H_2PO_4 + NH_4Cl$$

It has been shown that the most satisfactory results are obtained over the pH range 6.4 – 6.9 (optimum pH is 6.6), and that when the pH exceeds 7.0 the precipitate is difficult to be removed from the beaker. To obtain a suitable pH for precipitation, ammonium acetate and ammonium chloride are added. Methyl red indicator is utilized to adjust the suitable pH of the solution.

$$ZnNH_4PO_4 = 65.37 + 14 + 4 + 31 + 64$$
$$= 178.37$$

$\because$ 178.37 g of zinc ammonium phosphate contains 65.37 g of zinc

$\therefore$ 1 g of zinc ammonium phosphate contains $\dfrac{65.37}{178.37}$

$= 0.366$ g of zinc.

Procedure: First of all 25 c.c. of the given zinc salt solution is taken in a beaker and 100 c.c. of distilled water is added to it. Now 1-2 c.c. of conc. hydrochloric acid is added and then about 2 drops of methyl red are added. The colour of the solution becomes red. Now dilute ammonia solution is added dropwise till the solution becomes just yellow. Then 1 g of solid ammonium acetate (A.R.) and 5 g of ammonium chloride (A.R.) are added. The solution is heated almost to boiling and 50 c.c. of a freshly prepared 10% solution of diammonium hydrogen phosphate is added. The solution is maintained just below the boiling point for further 30 minutes to allow the precipitate to become granular and crystalline. The colour of the solution at this stage should be yellow, but not pink. Heating is stopped and the solution allowed to stand for 1-2 hours.

Now the supernatant liquid is transferred to a weighed (G-4) sintered crucible. The precipitate is washed 3-4 times with 1% diammonium hydrogen phosphate solution by decantation. Then the precipitate is transferred from the beaker to the crucible making use of a 'policeman'. The precipitate is washed with 1% diammonium hydrogen phosphate solution, until the washings are free from chloride ions. Finally, the precipitate is washed 3-4 times with 50% ethanol to remove excess of diammonium hydrogen phosphate from the precipitate. The precipitate is dried in an electric oven at 110–120°C for an hour. It is cooled and weighed till constant weight is obtained.

Calculation:

Wt. of sintered glass crucible = 24.576 g (say)

Wt. of sintered glass crucible + zinc ammonium phosphate

= 24.903 g (say)

$\therefore$ Wt. of zinc ammonium phosphate = 0.327 g

$\because$ 1 g of zinc ammonium phosphate contains 0.366 g of zinc

∴ 0.327 g of zinc ammonium phosphate contains 0.366 × 0.327 = 0.120 g of zinc

∵ 25 c.c. of zinc salt solution contains 0.120 g of zinc

∴ 1000 c.c. of zinc salt solution contains $\frac{0.120}{25} \times 1000$ = 4.8 g of zinc

Therefore, amount of zinc = 4.8 g/litre.

Things to Remember

(i) Preparation of zinc salt solution: 6 g of zinc oxide is taken in a measuring flask of 1000 c.c. 10–20 c.c. of conc. HCl (or, conc. HNO_3 or, dil. H_2SO_4) is added and shaken to dissolve. Now it is diluted to 1000 c.c. with distilled water.

or,

Zinc sulphate, $ZnSO_4$ (A.R.) crystals may be dissolved in distilled water.

(ii) Preparation of precipitation reagent (20% diammonium hydrogen phosphate solution): 20 g of diammonium hydrogen phosphate is dissolved in 100 c.c. of distilled water.

(iii) Preparation of wash solutions:

- 1% Diammonium hydrogen phosphate: 1 g of diammonium hydrogen phosphate is dissolved in 100 c.c. of distilled water.
- 50% Ethanol: 10 c.c. of ethanol is diluted with 10 c.c. of distilled water.

(iv) Preparation of ammonia solution: It is prepared by adding 1 c.c. of liquor ammonia in 20 c.c. of distilled water.

2.9. To estimate magnesium as magnesium pyrophosphate in the given solution of magnesium salt

Theory: Magnesium salt solution is treated with excess of diammonium hydrogen phosphate and ammonia solution to get a precipitate of magnesium ammonium phosphate hexahydrate which on ignition yields a precipitate of magnesium pyrophosphate. It is weighed.

$$MgSO_4 + (NH_4)_2 HPO_4 + NH_3 \rightarrow Mg(NH_4)PO_4 + (NH_4)_2 SO_4$$

$$2Mg(NH_4)PO_4 \rightarrow \underset{\text{(Mag. pyrophosphate)}}{Mg_2P_2O_7} + 2NH_3 + H_2O$$

$$Mg_2P_2O_7 = 48 + 62 + 112 = 222$$

$\because$ 222 g of mag. pyrophosphate contains 48 g of Mg

$\therefore$ 1 g of mag. pyrophosphate contains $\frac{48}{222}$ g of Mg

= 0.216 g of Mg

Procedure: First of all 25 c.c. of magnesium salt solution is taken in a beaker. 25 c.c. of distilled water, 2-3 c.c. of conc. HCl, 2 g of NH_4Cl (A.R.) and a few drops of methyl red are added one by one. Then 25% diammonium hydrogen phosphate solution is added in excess with constant stirring. Then 50% ammonia solution is added dropwise with constant stirring until the solution in the beaker turns yellow. 2-3 c.c. of ammonia solution is added in excess and the solution stirred for few minutes. While stirring one should not scratch the sides of the beaker with the stirring rod. The precipitate is allowed to settle for at least 4 hours or preferably over night.

The supernatant liquid is decanted to the Whatman filter paper No. 40. The precipitate is washed by decantation with 5% ammonia solution. After a few washings, the precipitate is transferred from the beaker to the filter paper and washing continued till the filtrate is free from chloride and sulphate ions.

After washing, the precipitate is dried in an electric oven. The precipitate is separated from the filter paper and the filter paper is ignited in a weighed crucible over a low flame. It is cooled and the precipitate is transferred to the crucible and heated strongly for an hour. It is cooled and weighed until constant weight is obtained.

Calculation:

Wt. of crucible = 30.217 g (say)

Wt. of crucible + mag. pyrophosphate = 30.516 g

$\therefore$ Wt. of $Mg_2P_2O_7$ = 0.299 g

$\because$ 1 g of $Mg_2P_2O_7$ contains 0.216 g of Mg

$\therefore$ 0.299 g of $Mg_2P_2O_7$ contains 0.216 × 0.299 g of Mg
= 0.0645 g of Mg

$\because$ 25 c.c. of Mg salt solution contains 0.0645 g of Mg

$\therefore$ 1000 c.c. of Mg salt solution contains $\frac{0.0645}{25} \times 1000$

$= \frac{64.5}{25}$

= 2.58 g of Mg

Therefore, amount of Mg = 2.58 g/litre.

Things to Remember

(i) Preparation of magnesium salt solution: 26.5 g of pure $MgSO_4 \cdot 7H_2O$ crystals is dissolved in 1000 c.c. of distilled water.

(ii) Preparation of precipitating reagent: 25 g of diammonium hydrogen phosphate is dissolved in 100 c.c. of distilled water.

(iii) Ammonia solution: It is 1:1 ammonia solution which may be prepared by adding 50 c.c. of liquor ammonia to 50 c.c. of distilled water.

(iv) Washing solution: It is 5% ammonia solution which is prepared by adding 5 c.c. of liquor ammonia to 95 c.c. of distilled water.

(v) The estimation of magnesium as mag. amm. phosphate hexahydrate may be done by above method. One should remember that the precipitate of mag. amm. phosphate should be filtered in a weighed sintered glass crucible and dried in air and desiccator.

(vi) The estimation of magnesium as mag. amm. phosphate does not give good result due to number of reasons.

(vii) The precipitate which has been allowed to stand overnight is less soluble than that which has been stood for four hours, and the precipitation is more complete.

(viii) It is advisable to cover the beaker while the precipitate is standing in order to minimize evaporation of ammonia.

(ix) The ignition of precipitate must be affected at a high temperature—about 1100°C.

(x) The ignition of filter paper must be carried out at low temperature.

(xi) Mag. pyrophosphate dissolves readily in nitric, hydrochloric, or acetic acid, which can be used for cleaning crucible after a determination.

2.10. To estimate calcium as calcium oxide in the given solution of calcium salt

Theory: Estimation of calcium is based on the precipitation of calcium as calcium oxalate by treating a hot HCl solution of the sample with ammonium oxalate solution, followed by neutralisation of the solution by means of ammonia. The precipitate, after being washed and dried, is ignited strongly, so as to convert it into the oxide which is weighed.

$$\begin{aligned} Ca^{2+} + C_2O_4^{2-} &\rightarrow CaC_2O_4 \\ CaC_2O_4 &\rightarrow CaCO_3 + CO \text{ (at dull redness)} \\ CaCO_3 &\rightarrow CaO + CO_2 \text{ (at about 1200°C)} \end{aligned}$$

$$\begin{aligned} CaO &= 40 + 16 \\ &= 56 \end{aligned}$$

$\because$ 56 g of CaO contains 40 g of Ca

$\therefore$ 1 g of CaO contains $\frac{40}{56}$ g of Ca

$= 0.714$ g of Ca

Procedure: First of all 25 c.c. of the given solution of calcium ion is taken in a beaker. About 50 c.c. of distilled water is added to it. 4-5 c.c. of hydrochloric acid (1:1) is added if it has not been added in the preparation of calcium salt solution. The solution is heated to boiling. To this hot solution, 5 g of ammonium chloride (A.R.) and 2 drops of methyl red are added. The solution is further heated to boiling and then 50 c.c. of a hot solution of ammonium oxalate is added. Now ammonia solution (1:1) is added with stirring until the colour just turns yellow. The solution is allowed to stand for one hour. The supernatant liquid is tested with a drop of ammonium oxalate solution to make sure that the precipitation is complete.

Clear supernatant liquid is decanted off through a gravimetric filter paper (Whatman No. 41). The precipitate in a beaker is washed 3-4 times by decantation with 0.2% solution of ammonium oxalate. Finally, the precipitate is transferred from the beaker to the filter paper by 0.2% solution of ammonium oxalate. Again the precipitate on the filter paper is washed with 0.2% solution of ammonium oxalate till the filtrate is free from chloride ions.

The precipitate is dried by keeping the funnel in an air oven at 120°C. The filter paper is folded so as to completely enclose the precipitate and placed in a weighed silica crucible. Now it is ignited for an hour. It is cooled and weighed till constant weight is obtained.

Calculation:

Wt. of silica crucible = 30.217 g (say)

Wt. of silica crucible + calcium oxide = 30.517 g (say)

$\therefore$ Wt. of calcium oxide = 0.3 g

$\because$ 1 g of CaO contains 0.714 g of Ca

$\therefore$ 0.3 g of CaO contains 0.714 × 0.3 g of Ca

= 0.214 g of Ca

$\because$ 25 c.c. of Ca-salt solution contains 0.214 g of Ca

$\therefore$ 1000 c.c. of Ca-salt solution contains $\frac{0.214}{25} \times 1000$ g

= 8.56 g of Ca

Therefore, amount of Ca = 8.56 g/litre.

Things to Remember

(i) Preparation of Ca-salt solution: 21.4 g of pure calcium carbonate is dissolved in dil. HCl and the volume is made 1000 c.c. with distilled water.

(ii) Precipitation solution: It is 4% ammonium oxalate which is prepared by dissolving 10 g of ammonium oxalate (A.R.) in 250 c.c. of distilled water.

(iii) Ammonia solution: It is 1:1 or 50% ammonia and is prepared by diluting 50 c.c. of liquor ammonia to 100 c.c. with distilled water.

(iv) Hydrochloric acid: It is 1:1 hydrochloric acid which is prepared by diluting 50 c.c. of conc. HCl to 100 c.c. with distilled water.

(v) Wash solution: It is 0.2% ammonium oxalate solution and is prepared by dissolving 0.2 g of ammonium oxalate in 100 c.c. of distilled water.

(vi) The residue of CaO is hygroscopic. It should, therefore, be cooled in desiccator and then weighed as early as possible.

6 ESTIMATION OF TWO CONSTITUENTS WHEN PRESENT TOGETHER

1. ANALYSIS OF GIVEN SOLUTION

1.1. To estimate copper and zinc in the given solution

Procedure: It involves the following steps:

(a) **Volumetric estimation of copper:** 25 c.c. of the given solution is pipetted out and the excess of acid neutralised by adding solid Na_2CO_3 till the effervescence ceases. The excess of Na_2CO_3 is neutralised by the addition of glacial acetic acid. Now the titration is carried out in the same manner as described in section 15.3 of Chapter 4.

(b) **Gravimetric estimation of copper:** 50 c.c. of the given solution is pipetted out and taken in a beaker. 2 c.c. of dilute hydrochloric acid is added to it. About 30-35 c.c. of a freshly prepared saturated solution of sulphurous acid is added. Smell of sulphur dioxide in the beaker indicates the presence of excess of sulphurous acid. The solution is diluted by adding 100 c.c. of distilled water. The solution is heated to boiling. Heating is stopped and then 50 c.c. of 10% ammonium thiocyanate solution is added with constant stirring. The precipitate is digested on a water bath and the content is tested for complete precipitation with 10% ammonium thiocyanate solution.

The solution is filtered through a weighed G-4 glass crucible and the precipitate washed with 1% ammonium thiocyanate solution. The filtrate and the washings are kept reserve for the estimation of zinc. The precipitate is dried by keeping the crucible at 110–120°C in air oven and weighed till constant weight (For details, see section 2.7 of Chapter 5).

(c) **Gravimetric estimation of zinc:** The filtrate and washings, obtained above are taken in a basin and evaporated to 80–100 c.c. To this 35 c.c. of conc. nitric acid and 15 c.c. of conc. HCl are added and the solution dried on a sand bath almost to dryness. This destroys the excess of thiocyanate added in estimation of copper. The residue is now dissolved in distilled water. Then the estimation of zinc is carried out as explained in section 2.8 of Chapter 5.

Calculation: See Chapters 4 and 5.

Result:

Copper (volumetrically) = ... g/litre

Copper (gravimetrically) = ... g/litre

Zinc (gravimetrically) = ... g/litre

Things to Remember

The solution containing copper and zinc ions can be prepared as follows:

- 30 g of copper sulphate is dissolved in distilled water in presence of glacial acetic acid and the volume is made 500 c.c.
- 8 g of zinc oxide is dissolved in distilled water containing sulphuric acid and the volume is made 500 c.c.
- 50 c.c. of each of the above two solutions are taken and mixed together. The volume is made 250 c.c. with the help of distilled water. Now the solution is ready to supply to the students.

1.2. To estimate silver and zinc in the given solution

Procedure: It involves the following steps:

(a) **Volumetric estimation of silver:** It is carried out in a similar manner as described in section 16.1 or 16.2 of Chapter 4 using standard solution of NaCl or KSCN respectively.

(b) Gravimetric estimation of silver: 50 c.c. of the supplied solution is taken in a beaker. 1 c.c. of conc. HNO_3 is added and diluted with 50 c.c. of distilled water. It is heated upto 70°C. To this hot solution, 0.2N HCl is added with constant stirring till the precipitation is complete. The precipitate is digested on a water bath for about 30–35 minutes. It is filtered through a G-3 crucible. The precipitate is washed with distilled water containing nitric acid until the washings are free from chloride ions. Finally the precipitate is washed with distilled water to remove nitric acid. The filtrate and washing are kept reserve for the gravimetric estimation of zinc (For details of estimation of silver, see section 2.3 of Chapter 5).

(c) Gravimetric estimation of zinc: The filtrate and washings are transferred to a basin and evaporated to 100 c.c. Then zinc is estimated from this solution as discussed in section 2.8 of Chapter 5.

Calculation: See Chapters 4 and 5.

Result:

Silver (volumetrically) = ... g/litre
Silver (gravimetrically) = ... g/litre
Zinc (gravimetrically) = ... g/litre

Things to Remember

Solution for such an exercise may be prepared as follows:

- 15 g of silver nitrate is dissolved in 500 c.c. of distilled water containing small amount of nitric acid.
- 8 g of zinc oxide is dissolved in 500 c.c. of distilled water containing small amount of nitric acid.
- 50 c.c. of each of the above solutions are mixed together and diluted to 250 c.c. with distilled water. The solution is ready for examination.

1.3. To estimate copper and silver in a given solution

Procedure: It involves the following steps:

(a) Volumetric estimation of silver: Volumetric estimation of silver can be carried out as described in section 16.1 or 16.2

of Chapter 4 using standard solution of NaCl or NaSCN respectively.

(b) Volumetric estimation of copper: 25 c.c. of the given solution is taken in a conical flask. To this excess of potassium iodide solution is added to precipitate whole of the silver as silver iodide.

$$AgNO_3 + KI \rightarrow AgI \downarrow + KNO_3$$

Rest of the procedure is as explained in section 15.3 of Chapter 4. It is better to remove Ag^+ ions from the solution.

(c) Gravimetric estimation of silver: Silver is estimated as silver chloride as discussed in section 2.3 of Chapter 5 taking 50 c.c. of the given solution. The filtrate and washings are kept reserve for the gravimetric estimation of copper.

(d) Gravimetric estimation of copper: The above filtrate and washings are taken in a basin and evaporated to almost dryness. Now 1-2 c.c. of conc. H_2SO_4 is added. Again it is heated till white dense fumes of sulphur dioxide are evolved. It is cooled and the sides of the basin washed down and transferred to a beaker. It is diluted to 100 c.c. with distilled water. The solution is neutralised with 1:1 ammonia solution. Then gravimetric estimation of copper is performed as described in section 2.7 of Chapter 5.

Calculation: See Chapters 4 and 5.

Result:

Silver (volumetrically) = ... g/litre

Copper (volumetrically) = ... g/litre

Silver (gravimetrically) = ... g/litre

Copper (gravimetrically) = ... g/litre

Things to Remember

Solution for such an exercise may be prepared as given below:

- 15 g of $AgNO_3$ is dissolved in distilled water containing nitric acid and the volume is made 500 c.c.

- 30 g of copper sulphate is dissolved in distilled water containing little acetic acid and the volume is made 500 c.c. with distilled water.
- 50 c.c. of each of the above solutions are mixed together and the volume is made 250 c.c. with distilled water. Now the solution is ready for examination.

1.4. To estimate silver and nickel in a given solution

Procedure: It involves the following steps:

(a) **Volumetric estimation of silver:** Silver is estimated volumetrically in a similar manner as described in section 16.1 or 16.2 of Chapter 4 using standard solution of NaCl or KSCN respectively.

(b) **Gravimetric estimation of silver:** Silver is estimated as silver chloride in 50 c.c. of the given solution. For procedure and other details section 2.3 of Chapter 5 can be seen. The filtrate and washings are kept reserve for the gravimetric estimation of nickel.

(c) **Gravimetric estimation of nickel:** The filtrate and washings are taken in a basin and evaporated to 50 c.c. The sides of the basin are washed down and transferred to a beaker. Now nickel is estimated as nickel dimethylglyoximate in a similar manner as described in section 2.6 of Chapter 5.

Calculation: See Chapters 4 and 5.

Result:

Silver (volumetrically) = ... g/litre

Silver (gravimetrically) = ... g/litre

Nickel (gravimetrically) = ... g/litre

Things to Remember

The solution for such an exercise may be prepared as follows:

- 15 g of silver nitrate is dissolved in distilled water containing nitric acid and the volume is made 500 c.c.

- 20 g of nickel ammonium sulphate is dissolved in distilled water and the volume is made 500 c.c.
- 50 c.c. of each of the above two solutions are taken in a 250 c.c. measuring flask and diluted upto the mark with distilled water. Now the solution is ready for examination.

1.5. To estimate copper and magnesium in the given solution

Procedure: It involves the following steps:

(a) **Volumetric estimation of copper:** Volumetric estimation of copper is done taking 25 c.c. of the given solution (For details, see section 15.3 of Chapter 4).

(b) **Gravimetric estimation of copper:** 50 c.c. of the given solution is taken and copper is estimated gravimetrically as described in section 2.7 of Chapter 5. The filtrate and washings are kept reserve for the estimation of magnesium.

(c) **Gravimetric estimation of magnesium:** The filtrate and washings are taken in a basin and evaporated to 80–100 c.c. Now 25 c.c. of conc. HNO_3 and 15 c.c. of conc. HCl are added and the solution evaporated almost to dryness. Finally, the solution is evaporated to dryness to expel excess of HNO_3. Sides of the basin are washed down and then transferred to a beaker. Now magnesium is estimated in a similar manner as described in section 2.9 of Chapter 5.

Calculation: See Chapters 4 and 5.

Result:

Copper (volumetrically) = ... g/litre

Copper (gravimetrically) = ... g/litre

Magnesium (gravimetrically) = ... g/litre

Things to Remember

The solution for such an exercise may be prepared as follows:

- 30 g of copper sulphate is dissolved in distilled water

in presence of glacial acetic acid and the volume is made 500 c.c. with distilled water.

- 30 g of magnesium sulphate is dissolved in distilled water and the volume is made 500 c.c.
- 50 c.c. of each of the above solutions are mixed together and the volume is made 250 c.c. Now the solution is ready for examination.

1.6. To estimate iron and nickel in a given solution

Procedure: It involves the following steps:

(a) **Volumetric estimation of iron:** Volumetric estimation of iron is carried out in a similar manner as described in section 14.3 of Chapter 4. Standard $KMnO_4$ solution should be used to titrate the solution of iron of unknown strength.

(b) **Gravimetric estimation of iron:** 50 c.c. of the given solution is pipetted out and taken in a beaker. To this, 5 g of solid ammonium chloride is added to prevent the precipitation of nickel hydroxide along with iron hydroxide. Then the iron is estimated as ferric oxide in a similar manner as described in section 2.5 of Chapter 5. The filtrate and washings are kept reserve for the gravimetric estimation of nickel.

(c) **Gravimetric estimation of nickel:** The filtrate and washings are evaporated to 100 c.c. and then nickel is estimated as nickel dimethylglyoximate as described in section 2.6 of Chapter 5.

Calculation: See Chapters 4 and 5.

Result:

Iron (volumetrically) = ... g/litre

Iron (gravimetrically) = ... g/litre

Nickel (gravimetrically) = ... g/litre

Things to Remember

The solution for such an exercise may be prepared as follows:

- 20 g of nickel ammonium sulphate is dissolved in distilled water in presence of sulphuric acid and the volume is made 500 c.c. with the help of distilled water.

- 50 g of ferrous ammonium sulphate is dissolved in distilled water in presence of sulphuric acid and diluted to 500 c.c. with distilled water.
- 50 c.c. of each of the above solutions are mixed together and diluted to 250 c.c. with distilled water. Now the solution is ready for analysis.

1.7. To estimate copper and nickel in a given solution

Procedure: It involves the following steps:

(a) **Volumetric estimation of copper:** 25 c.c. of the given solution is taken to estimate copper volumetrically. See the details in section 15.3 of Chapter 4.

(b) **Gravimetric estimation of copper:** 50 c.c. of the given solution is taken in a beaker. 2-3 c.c. of hydrochloric acid and 30-40 c.c. of a freshly prepared saturated solution of sulphurous acid are added to it. Smell of SO_2 in the beaker indicates the presence of excess of sulphurous acid. The solution is diluted by adding 100 c.c. of distilled water. The solution is heated to boiling. Heating is discontinued and then 10% ammonium thiocyanate solution (25-30 c.c.) is added slowly with constant stirring. The precipitate is digested by keeping the beaker on a water bath for about 30 minutes and the supernatant liquid is tested for complete precipitation by adding one or two drops of 10% ammonium thiocyanate solution. The solution is then filtered through a weighed sintered G-4 glass crucible. The precipitate is washed with 1% ammonium thiocyanate solution. The filtrate and washings are kept reserve for the estimation of nickel. The precipitate is dried by keeping the crucible in an electric oven at 110-120°C and then the crucible is weighed till constant weight is obtained.

(c) **Gravimetric estimation of nickel:** The filtrate and washings obtained from above are taken in a basin and evaporated to 80-100 c.c. Now 35 c.c. of conc. HNO_3 and 15 c.c. of conc. HCl are added to it and then the solution is evaporated nearly to dryness (3-4 c.c.) on a sand bath. Sides of the basin are washed down with distilled water and then the solution is evaporated to complete dryness in order to remove excess of nitric acid. The residue is dissolved in water and 1 c.c. of conc. HCl is added (if the solution is not clear). The solution is transferred from the basin to the beaker. The solution is diluted to 150 c.c. with distilled water. 2 drops of methyl red are added. Then 30 c.c. of 1% dimethylglyoxime is added with constant stirring. Now 1:1 ammonia solution is added dropwise till the solution becomes alkaline, i.e., the colour of the solution changes from red to yellow. The precipitate is digested on a water bath and the test for complete precipitation is done by adding few drops of dimethylglyoxime to the clear supernatant liquid. The solution is cooled for one hour and filtered through a weighed G-3 crucible. The precipitate is washed many times with hot water until the washings are free from Cl^- and SO_4^{2-} ions. Now the precipitate is dried at 110-120° by keeping the crucible in an electric oven and weighed till constant weight is obtained.

Calculation: See Chapters 4 and 5.

Result:

Copper (volumetrically) = ... g/litre

Copper (gravimetrically) = ... g/litre

Nickel (gravimetrically) = ... g/litre

Things to Remember

Solution for such an exercise may be prepared as follows:

- 30 g of copper sulphate is dissolved in distilled water in presence of glacial acetic acid and the volume is made 500 c.c.
- 20 g of nickel ammonium sulphate is dissolved in distilled water and the volume is made 500 c.c. by diluting it with distilled water.
- 50 c.c. of each of the above solutions are mixed together and diluted to 250 c.c. with distilled water.

1.8. To estimate iron and magnesium in the given solution

Procedure: It involves the following steps:

(a) Volumetric estimation of iron: 25 c.c. of the given solution is taken in a conical flask and titrated against standard $KMnO_4$ solution. For details, see section 14.3 of Chapter 4.

(b) Gravimetric estimation of iron: 50 c.c. of the given solution is taken and then iron is estimated as ferric oxide in a similar manner as described in section 2.5 of Chapter 5. The filtrate and washings are kept reserve for the gravimetric estimation of magnesium.

(c) Gravimetric estimation of magnesium: The filtrate and washings obtained from above are taken in a basin and evaporated to 25 c.c. It is transferred to a beaker completely and then magnesium is estimated following the same procedure as described in section 2.9 of Chapter 5.

Calculation: See Chapters 4 and 5.

Result:

Iron (volumetrically) = ... g/litre

Iron (gravimetrically) = ... g/litre

Magnesium (gravimetrically) = ... g/litre

2. ANALYSIS OF ALLOYS

When an ore or alloy is taken for the quantitative analysis, it is essential that the various constituents present in the sample be known. The percentage of constituents may vary from sample to sample. Some elements are in large quantity, whereas a few elements may be present as trace constituents. Hence, the sample should be analysed qualitatively first to find out the presence of the elements as constituents.

Once the constitutents are known, then separation and subsequent estimation of individual constitutents is carried out. Each element of the constituent is removed from the solution according to its placement in a particular group, e.g., copper is always removed as CuS while estimating aluminium, iron, zinc, manganese, nickel, barium, etc., which are placed in Group III onwards in qualitative analysis.

Analysis of ores: The sample of ore is crushed to fine powder in a mortar. The powdered ore is weighed on a dry watch glass and then heated in an oven at 110°C or in a crucible. Thus moisture and volatile impurities are removed and a stable dried sample is obtained. It is weighed again to find out the loss on heating.

Then the dried sample is heated strongly with 1:1 or conc. HCl in a porcelain basin or a beaker covered with a watch glass.

It is evaporated almost to dryness, residue is cooled and extracted again with dil. HCl to obtain a solution. It is filtered if required. If the insoluble residue is found, then it is washed with dil. HCl and water; and ignited to get insoluble SiO_2, etc., and is weighed as such.

Filtrate is preserved and used for quantitative estimation of constituents.

Analysis of alloys: Alloys are generally soluble in conc. HNO_3. They are placed in a beaker or in a porcelain basin and dissolved in conc. HNO_3. Solution is evaporated almost to dryness, residue is cooled and then, extracted with dil. H_2SO_4, heated to remove traces of HNO_3.

The solution obtained is filtered, if insoluble matter is found to be present. This residue may be metastannic acid, is washed and finally ignited to obtain SnO_2 and weighed if required.

Filtrate is preserved and used for estimating the constituents present in alloy.

2.1. Analysis of Brass

The major constituents of brass are copper (50-70%) and zinc (20-40%) but small amount of tin (0-60%), lead (0-20%) and iron (0-1%) and sometimes traces of aluminium, nickel and manganese may also be present. The analysis of brass involves separation and determination of Cu, Zn, Sn, Pb and Fe.

Procedure: It involves following steps:

(a) Tin: About 1 g of the brass is weighed accurately and taken in a beaker. 15 c.c. of 10N HNO_3 is added to it and the beaker covered with a watch glass. It is evaporated slowly on a water bath to a volume of 5 to 10 c.c. but not to dryness. Now the digestion is carried out for atleast one hour to ensure complete precipitation of all the stannic acid. It is diluted with water to 50 c.c. and heated on the water bath for 25 minutes. This helps the soluble salt to dissolve. Some pieces of paper are added and stirred. It is filtered through Whatman filter paper No. 42. The precipitate is washed 10 times with hot dil. HNO_3 (1:100). The filtrate and washings are preserved for the determination of other constitutents. The precipitate is transferred in a weighed silica crucible. It is ignited at 900-1000°C till constant weight of SnO_2 is obtained.

Generally result of tin by the above method is slightly high due to the presence of small quantities of Fe, Cu, Zn and Pb in the ignited residue. Therefore, a correction is made in the following way:

> To the constant weight of impure SnO_2, fifteen times of NH_4I to that of SnO_2 is added and heated at about 45°C until the fumes stop to come off. Tin is volatilised as SnI_4 but other metals do not. It is cooled and weighed again. The difference between the two weights gives the weight of pure SnO_2.

(b) Lead: The filtrate and washings obtained from above are concentrated. It is cooled and then 3-4 c.c. of conc. H_2SO_4 is added. The solution is heated with stirring until heavy white fumes are evolved. Heating is continued to ensure complete expulsion of all HNO_3. It is cooled and diluted to 100 c.c. Further it is heated nearly to boiling and then kept to cool for one hour. The precipitate ($PbSO_4$) is filtered through a weighed sintered crucible and washed thoroughly with 1% H_2SO_4. The filtrate and washings are kept reserve for the determination of other constituents. The precipitate is dried at 130°C to constant weight.

(c) **Copper:** The filtrate and washings are evaporated to 50 c.c. It is neutralised with 1:1 ammonia solution. Copper is then precipitated as CuSCN. The filtrate and washings are kept reserve for the estimation of iron and zinc.

(d) **Iron:** To this filtrate and washings, 35 c.c. of conc. HNO_3 and 15 c.c. of conc. HCl are added. The solution is evaporated nearly to dryness. It is cooled. Further 3-5 c.c. of conc. HCl is added and warmed. It is diluted to about 50-75 c.c. The iron is then precipitated as $Fe(OH)_3$. It is filtered. The filtrate and washings are kept reserve for zinc. Iron is weighed as iron oxide.

(e) **Zinc:** Zinc is estimated from the filtrate and washings as zinc ammonium phosphate by adding diammonium hydrogen phosphate at pH of 5.5 to 7 maintained by methyl red indicator (as described in section 2.8 of Chapter 5).

Result:

Tin = ... %
Lead = ... %
Copper = ... %
Iron = ... %
Zinc = ... %

2.2 Analysis of Solder

Commercial solder consists essentially of lead and tin, but it also contains a very small amount of antimony (1-4%).

Procedure: About 0.5 g of the finely divided alloy is weighed accurately and taken in a beaker. 5 c.c. of distilled water and 15 c.c. of concentrated nitric acid are added one by one. The beaker is covered with watch glass during the violent reaction. When the vigorous reaction is over, it is evaporated on a water bath to a volume of about 5 c.c., but not to dryness. The solution is diluted to 50 c.c. and heated on a water bath for 15 minutes with stirring. Lead dissolves, while tin and antimony remain in insoluble conditions as stannic and antimonic acids (H_2SnO_3, H_3SbO_4), respectively along with small quantity of co-precipitated PbO. The precipitate is filtered through a Whatman filter paper number 42. The precipitate is washed at least ten times with hot water containing dilute nitric acid. The filtrate and washings are kept reserve to estimate lead as $PbSO_4$.

$$3Sn + H_2O + 4HNO_3 \rightarrow 3H_2SnO_3 + 4NO\uparrow$$

$$3Sb + 2H_2O + 5HNO_3 \rightarrow 3H_3SbO_4 + 5NO\uparrow$$

(a) **Tin:** The precipitate of mixed acids is dried as usual. It is ignited with filter paper in a weighed crucible for half an hour. The crucible is allowed to cool and weighed as mixed oxides (SnO_2 + Sb_2O_4 + PbO).

$$H_2SnO_3 \rightarrow SnO_2\downarrow + H_2O$$

$$4H_3SbO_4 \rightarrow 2Sb_2O_4\downarrow + 6H_2O + O_2\uparrow$$

Fifteen times of ammonium iodide (powder) to that of mixed oxides is added and mixed thoroughly. It is heated at about 450°C until fumes stop to evolve. Here tin and antimony are volatilised as their iodides. Now the crucible is cooled and then 2-3 c.c. of conc. HNO_3 is added. It is evaporated to dryness. Now it is ignited to constant weight. The loss in weight gives the weight of SnO_2 + Sb_2O_4. Percentage of antimony may be calculated separately by the method given in the last para of this experiment. Hence percentage of tin can be easily calculated.

(b) **Lead:** The residue in the crucible is dissolved in a few c.c. of conc. HCl. This solution is added to the filtrate and washings as obtained earlier. 10 c.c. of conc. H_2SO_4 is added and then the solution is boiled. It is cooled to room temperature and then 5 c.c. of rectified spirit is added. The function of rectified spirit (ethanol) is to reduce the solubility of $PbSO_4$. The precipitate is allowed to stand for half an hour. The supernatant liquid is tested for complete precipitation by adding a few drops of dil. H_2SO_4. The precipitate of $PbSO_4$ is filtered through a weighed sintered glass crucible G-3. The precipitate is washed 3 to 4 times with 1% sulphuric acid. It is dried at 130°C and weighed to constant weight.

(c) **Antimony:** Antimony may be estimated in the alloy volumetrically by titrating the solution against standard $KMnO_4$ solution which oxidises $SbCl_3$ to $SbCl_5$.

$$2KMnO_4 + 5SbCl_3 + 16HCl \rightarrow 5SbCl_5 + 2KCl + 2MnCl_2 + 8H_2O$$

Tin does not interfere in the solution and lead is removed from the solution since it may absorb some antimony. 0.5 g of the alloy is weighed accurately and dissolved in 25 c.c. of conc. H_2SO_4 in a pyrex conical flask or beaker. The flask is heated for sometime until the alloy is completely disintegrated. The solution is allowed to cooled and diluted very carefully with 100 c.c. of distilled water. The precipitate of $PbSO_4$ is allowed to settle down. It is filtered.

The precipitate is warmed with 20 c.c. of conc. HCl and 10 c.c. of water and then it is diluted to 100 c.c. It is boiled to expel SO_2. Now this solution is added to the main solution (filtrate) containing antimony and the volume made up to 250 c.c. with water. It is titrated with 0.1N $KMnO_4$ very slowly to avoid the oxidation of HCl. From the volume of $KMnO_4$ used, the amount of antimony present in the solution may be calculated.

Result:

Tin = ... %

Lead = ... %

Antimony = ... %

APPENDIX

1. CONCENTRATED REAGENTS

Name	Sp. Gravity	Approx. Normality
Acetic acid glacial	1.05	17 N
Acid hydrochloric	1.19	12 N
Acid nitric	1.42	16 N
Acid sulphuric	1.84	36 N
Liquor ammonia	0.88	15 N

Things to Remember

Concentrated acetic acid is known as glacial acetic acid. When cooling to 0°C, the reagent freezes forming ice-like crystals, hence the name 'glacial'. It has to be handled with care.

2. PREPARATION OF SOLUTIONS ON STUDENTS' SHELF

Reagent	Preparation	Approx. Normality
Acetic acid	Mix glacial acetic acid = 235 c.c. distilled water = 1765 c.c.	2N
Hydrochloric acid	Mix conc. HCl = 333 c.c. distilled water = 1667 c.c.	2N
Nitric acid	Mix conc. HNO_3 = 250 c.c. distilled water = 1750 c.c. Add the acid very slowly to water.	2N

(*Contd...*)

Reagent	Preparation	Approx. Normality
Sulphuric acid	Mix conc. H_2SO_4 = 112 c.c. distilled water = 1888 c.c. Add the acid very slowly to water.	2N
Ammonium carbonate	Dissolve 192 g of the salt in a mixture of 140 c.c. liquor ammonia and 1860 c.c. of distilled water.	2N
Ammonium hydroxide	Mix liquor ammonia = 267 c.c. distilled water = 1773 c.c.	2N
Ammonium sulphate	Dissolve 264 g of the salt in 2 litres of water.	2N
Barium chloride	Dissolve 122 g of $BaCl_2{\cdot}2H_2O$ in 2 litres of distilled water	0.5N
Cobalt nitrate	Dissolve 88 g of $Co(NO_3)_2{\cdot}6H_2O$ in 2 litres of distilled water.	0.3N
Sodium hydroxide	Dissolve 160 g of caustic soda in 2 litres of distilled water.	2N
Sodium carbonate	Dissolve 212 g of anhydrous salt in 2 litres of distilled water.	2N
Silver nitrate	Dissolve 34 g of the salt in 2 litres of distilled water.	0.1N

3. LIST OF REAGENTS ON STUDENTS' SHELF

Liquid

1. Acid acetic dilute
2. Acid hydrochloric dilute
3. Acid nitric dilute
4. Acid sulphuric dilute
5. Ammonium carbonate solution
6. Ammonium hydroxide solution
7. Ammonium sulphate solution
8. Barium chloride solution
9. Cobalt nitrate solution
10. Sodium hydroxide solution

11. Sodium carbonate solution
12. Methyl orange solution
13. Phenolphthalein solution
14. Silver nitrate solution

Solid

15. Ammonium chloride
16. Borax
17. Fusion mixture

4. PREPARATION OF SOLUTIONS ON SIDE SHELF

- Ammonium acetate — Dissolve 231 g of the salt in one litre of water.
- Ammonium molybdate — Dissolve 45 g of pure commercial ammonium molybdate in a mixture of 40 c.c. of liquor ammonia and 60 c.c. of water, and 120 g of ammonium nitrate, and dilute to one litre with water.
- Ammonium oxalate — Dissolve 35 g of the crystalline salt in one litre of water.
- Ammonium phosphate — Dissolve 74.5 g of the salt in one litre of water.
- Ammonium sulphide yellow — Saturate 200 c.c. liquor ammonia with H_2S, add 10 g of flowers of sulphur and 200 c.c. of liquor ammonia. Shake till the sulphur is dissolved and dilute to one litre.
- Ammonium thiocyanate — Dissolve 38 g of the salt in one litre of water.
- Barium chloride (0.25M) — Dissolve 61.1 g barium chloride dihydrate in water and dilute to 1 litre.
- Calcium chloride — Dissolve 55 g of the hydrated salt ($CaCl_2 \cdot 6H_2O$) in one litre of water.
- Calcium sulphate — Shake 3 g of the salt ($CaSO_4 \cdot 2H_2O$) with one litre of water; allow to stand; filter the saturated solution.

- Chlorine water — Saturate water with chlorine gas. The chlorine may be generated by dropping conc. HCl upon solid $KMnO_4$.
- Copper sulphate — Dissolve 125 g of the hydrated salt in one litre of water containing about 10 c.c. of dil. H_2SO_4.
- Dimethylglyoxime — Dissolve 1 g of dimethyl glyoxime in 125 c.c. rectified spirit; dilute it to 250 c.c. with distilled water.
- 2:4-Dinitrophenylhydrazine — Dissolve 1 g 2:4-dinitrophenylhydrazine in about 2 c.c. conc. HCl and dilute it with about 20 c.c. ethyl alcohol and further dilute it with about 100 c.c. dilute (2N) HCl solution.
- Fehling's Solution

  Solution No. 1 (Blue) — Dissolve 34.6 g of pure $CuSO_4 \cdot 5H_2O$ in water and dilute to 500 c.c.

  Solution No. 2 (colourless) — Dissolve 175 g of sodium potassium tartarate (Rochelle's salt) and 70 g of pure NaOH in water and dilute to 500 c.c.
- Ferric chloride — Dissolve 135 g of the hydrated salt, $FeCl_3 \cdot 6H_2O$ in one litre of water containing 20 c.c. concentrated hydrochloric acid.
- Iodine solution — Dissolve 12.7 g of iodine in a solution of 20 g of pure KI in 30 c.c. of water and dilute to one litre with water.
- Lead acetate — Dissolve 95 g of $Pb(CH_3COO)_2 \cdot 3H_2O$ in one litre of boiling water, cool. If the solution is turbid, add dil. acetic acid slowly till turbidity disappears.
- Mercuric chloride — Dissolve 27 g of the salt in one litre of water.
- Nessler's reagent — Dissolve 10 g of KI in 10 c.c. of water, add saturated mercuric chloride solution (60 g per litre) in small quantities at a time with shaking till a slight permanent precipitate is formed. Then

add 80 c.c. of 9M (67%) potassium hydroxide solution and then dilute to 200 c.c. Allow to stand overnight and decant the clear liquid. The reagent thus consists of an alkaline solution of potassium mercuric iodide, K_2HgI_4.

- α-Nitroso-β-naphthol — Dissolve 2.5 g α-nitroso-β-naphthol in 150 c.c. of glacial acetic acid, dilute to 250 c.c. with distilled water.
- Potassium dichromate — Dissolve 4.9 g of the salt in one litre of water.
- Potassium dihydrogen antimonate — The reagent (KH_2SbO_4) is prepared by adding 5.5 g potassium antimonate to 250 c.c. boiling water, boiling till nearly all the salt has dissolved, cooling rapidly, adding 9 c.c. 10% KOH solution, and filtering after standing overnight.
- Potassium ferricyanide — Dissolve 55 g of the salt in one litre water.
- Potassium ferrocyanide — Dissolve 53 g of the salt in one litre of boiling water.
- Potassium iodide — Dissolve 16.6 g of the salt in one litre of water.
- Potassium permanganate — Dissolve 3.2 g of the salt in one litre of water.
- Potassium thiocyanate — Dissolve 1 g pot. thiocyanate in 10 c.c. water.
- Sodium cobaltinitrite — Dissolve 17 g of the pure salt, $Na_3[Co(NO_2)_6]$ in 250 c.c. distilled water.
- Sodium nitroprusside — Dissolve 3-4 crystals in about 250 c.c. water by shaking.
- Stannous chloride — Boil a few pieces of tin metal with about 75 c.c. conc. HCl, cool and dilute to 250 c.c. Keep a few pieces of tin in the bottle to prevent oxidation.

- Starch solution — Grind about 1 g of soluble starch with a little cold water into a thin paste. Pour this paste into one litre boiling water, and cool.

5. LIST OF REAGENTS ON SIDE SHELF

Liquid

1. Acid acetic glacial
2. Acid hydrochloric concentrated
3. Acid nitric concentrated
4. Acid sulphuric concentrated
5. Ammonium acetate solution
6. Ammonium molybdate reagent
7. Ammonium molybdate 10% solution
8. Ammonium oxalate solution
9. Ammonium phosphate solution
10. Ammonium sulphide yellow solution
11. Ammonium thiocyanate solution
12. Bromine liquid
13. Calcium chloride
14. Calcium sulphate
15. Carbon tetrachloride
16. Chlorine water
17. Copper sulphate solution
18. Dimethylglyoxime solution
19. 2:4-Dinitrophenylhydrazine solution
20. Fehling's solution No. 1
21. Fehling's solution No. 2
22. Ferric chloride solution
23. Hydrogen peroxide (10 vol.)
24. Iodine in KI solution
25. Lead acetate solution
26. Liquor ammonia

27. Lime water
28. Mercuric chloride solution
29. Methyl alcohol
30. Methyl red indicator
31. Nessler's reagent solution
32. α-Nitroso-β-naphthol solution
33. Potassium chromate solution
34. Potassium dichromate solution
35. Potassium dihyrogen antimonate solution
36. Potassium ferricyanide solution
37. Potassium ferrocyanide solution
38. Potassium iodide solution
39. Potassium permanganate solution
40. Potassium thiocyanate solution
41. Schiff's reagent
42. Sodium cobaltinitrite solution
43. Sodium nitroprusside solution
44. Spirit rectified
45. Stannous chloride solution
46. Starch solution

Solid

47. Acid oxalic
48. Ammonium carbonate
49. Calcium fluoride
50. Charcoal animal
51. Copper foil
52. Copper turnings
53. Ferrous sulphate
54. Iron powder
55. Litmus paper (blue)
56. Litmus paper (red)

57. Manganese dioxide
58. Microcosmic salt
59. β-Naphthol
60. Potassium chromate
61. Potassium dichromate
62. Sodium acetate
63. Sodium bismuthate
64. Sodium bicarbonate
65. Sodium carbonate
66. Sodium nitrite
67. Sodium peroxide
68. Starch soluble
69. Tin metal
70. Zinc metal (granular)
71. Zinc nitrate

6. PREPARATION OF INDICATOR SOLUTION

(i) Dichlorofluorescein

It is 0.1% solution which is prepared by dissolving 0.1 g of its sodium salt in 100 c.c. of distilled water. 10 drops of this solution are sufficient for 25 c.c. of chloride solution for titrating it with $AgNO_3$ solution.

(ii) Di-iodomethylfluorescein

Its 1% solution is prepared by dissolving 1 g of it in 100 c.c. of 70% alcohol.

(iii) Diphenylamine

0.1 g of the diphenylamine is dissolved in 10 c.c. of concentrated H_2SO_4 to get 1% solution.

(iv) Eosin

Its 0.1% sodium salt solution is used which is prepared by dissolving 0.1 g of its sodium salt in 100 c.c. of distilled water. 10 drops of this solution are added to the 25 c.c. of bromide or iodide solution (when titrated with $AgNO_3$) which is diluted to 100 c.c.

(v) Eriochrome Black T

Eriochrome Black T is also known as Solochrome Black T. It is prepared by dissolving 0.2 g of it in 15 c.c. of triethanol amine and adding 5 c.c. of absolute ethanol to reduce the viscosity of the solution.

(vi) Ferric Alum

40 g of ferric alum is dissolved in 100 c.c. of distilled water to get 40% solution of it.

(vii) Litmus Solution (blue)

10 g of it is dissolved in a litre of water to get 1% of solution.

(viii) Methyl Orange

1 g of methyl orange is dissolved in a litre of water to get 0.1% solution.

(ix) Methyl Red

1 g of methyl red is dissolved in 1 litre of hot water, or 1 g of methyl red is dissolved in 600 c.c. of ethanol and diluted with 400 c.c. of water.

(x) Murexide

Aqueous solution of murexide is not stable for a long time so freshly prepared solution should be employed every day.

(xi) N-phenylanthranilic Acid

0.25 g of it is dissolved in 12 c.c. of 0.1N NaOH solution and diluted to 250 c.c. of water.

(xii) Patton and Reeddr's indicator

1 g of it is mixed with 100 g of A.R. anhydrous sod. sulphate in a glass mortar and 1 g of the mixture is used in each titration.

(xiii) Phenolphthalein

1 g of phenolphthalein is dissolved in 100 c.c. of alcohol to get 1% solution.

(xiv) Potassium Chromate

5 g of it is dissolved in 100 c.c. of water to get 5% solution.

(xv) Potassium Ferricyanide

1 g of it is dissolved in a litre of water to get 0.1% solution.

(xvi) Sodium Diphenylamine Sulphonate

0.2 g of it is dissolved in 100 c.c. of distilled water. Instead of this indicator, 0.2% aqueous barium diphenylamine sulphonate may frequently be used.

(xvii) Starch Solution

A paste of 1 g of starch with cold water is made. It is poured dropwise into about 100 c.c. of boiling water and continued to boil for few minutes. It is allowed to stand until cooled and then the clear solution is decanted.

If 0.5 g of salicylic acid is added to 100 c.c. of the starch solution then it keeps well for a long time.

7. CLEANING OF VOLUMETRIC GLASSWARE

Quantitative analysis demands proper cleaning of the apparatus. All glass apparatus must be scrupulously cleaned in order to obtain accurate results. If water does not wet surface of a vessel uniformly or the meniscus of a solution is distorted, it shows that apparatus is not clean. The contamination of glass surfaces with dirt, dust, grease, carbonaceous water or other substances of organic or inorganic nature modifies the surface tension of water or solution and thus causes error in the volume of the delivered. Some cleansing agents are listed as follows:

(i) Soap Solution: It may be used to remove ordinary contaminations.

(ii) Detergent Solution: It may be used to remove ordinary contaminations.

(iii) Sodium Hydroxide Solution (10%): It is used to remove carbonaceous matter.

(iv) Mixture of Potassium Dichromate and Concentrated Sulphuric Acid: 15-20 g of $K_2Cr_2O_7$ should be dissolved in 500 c.c. of cold and concentrated sulphuric acid and then this mixture may be used to remove grease, dust, and other contaminations.

(v) Solution of KOH in Ethanol: It is better than dichromate-sulphuric acid mixture.

(vi) Solution containing Na_3PO_4 (6%) and Sodium Oleate (3%): It is used to remove thick carbon deposit.

(vii) Concentrated Hydrochloric Acid: It is used to remove iron salts.

(viii) Mixture of Conc. HNO_3 and Conc. H_2SO_4: It may be used to remove carbonaceous matter.

(ix) Solution containing $FeSO_4$ and H_2SO_4: It may be used to remove $KMnO_4$ stains.

(x) Sodium Thiosulphate Solution: It may be used to remove iodine stains.

(xi) Alkaline $KMnO_4$ Solution: It may be used to remove grease. Final washing with conc. HCl and water is necessary.

(xii) Hot Decalin (Decahydronaphthalene): It may be used to remove silicon contaminations.

8. LIST OF HAZARDOUS CHEMICALS AND THEIR EFFECTS

Hazardous Chemical	Effect
(i) Salts of Ag, As, Ba, Cu, Hg, Ni, Pb, Sb, Tl, V, $C_2O_4^{2-}$, F^-, MnO_4^-	Most of these are very dangerous but only if swallowed. $AgNO_3$ causes caustic burn.
(ii) H_2S	Almost as poisonous as HCN. Exposure dulls the sense of smell.

(Contd...)

Hazardous Chemical	Effect
(iii) SO_2, NO_2, Cl_2, Br_2, I_2, HNO_3, H_2SO_4, HF	All are dangerous as well as unpleasant. When concentrated, all cause rapid destruction of the skin. HF is especially dangerous.
(iv) $HClO_3$, HClO, and their salts	Highly oxidising
(v) Chlorinated alkanes, e.g., $CHCl_3$, CCl_4	Most of these are narcotic, causing mental confusion.
(vi) Benzene	Toxic vapours causing dizziness.
(vii) Benzoyl chloride	Irritant.
(viii) Ether, ethanol	Very highly inflammable.
(ix) Nitrobenzene	Toxic vapours.
(x) Phenol	Burns the skin.

9. SAFETY MEASURES IN LABORATORY

An important aspect in a chemistry laboratory is your own and your fellow worker's safety. Accidents occur in the laboratory because of carelessness and inadequate knowledge about the chemicals being used. Though accidents cannot be fully eliminated, yet these can be prevented to some extent by knowing in advance some general precautionary measures. The following do's and don'ts in the laboratory would help you to avoid accidents.

(i) The Do's

- Wear a lab coat or an apron when working in the lab.
- Keep the test tube pointing away from yourself and others while heating it on a burner.
- Use splinters and not a paper to light a burner.
- You should know where the fire extinguishers are located in the laboratory and how to use them.
- Always use safety goggles for protecting your eyes from a dangerous operation, e.g., distillation of an inflammable liquid or while doing sodium ignition test.
- Wash your hands with soap when you leave the laboratory after doing an experiment.

- Carry out the reactions involving pungent or noxious fumes under a fume hood.
- Ensure that gas and water taps are closed before leaving the lab.

(ii) The Don'ts

- Don't wear loose clothes specially the synthetic ones while working in the laboratory.
- Don't taste any chemical, not even sucrose; it may be contaminated.
- Don't pipette out corrosive liquid by sucking with your mouth.
- Don't put the reagents back into the bottles or packs after use. These should be poured into another glass bottle kept specially for the waste liquids.
- Don't try to insert glass tubing or thermometer into corks forcibly.
- Don't inhale the vapours of any chemical deeply which might cause suffocation and choking; be alert and quick in perceiving the smell of the vapours, keeping the test tube in a slanting position.
- Don't keep inflammable solvents like petrol, ether, alcohol, etc. near a burner.
- Don't add pumice stones to a boiling liquid; add them before beginning to heat the liquid.
- Don't ever perform unauthorised experiments and never work alone in the laboratory.
- Don't touch electric switches with wet hands.

However, even if you are a careful worker and follow the general rules of safety, the accidents can occur—that's why they are called accidents.

10. FIRST-AID

In case of accident, you must be fully equipped and must know what to do in such a case. There should be a first-aid box in every laboratory containing some common things like

Dettol, Burnol, Band-aid, Bandages, Cotton, etc. Generally, the most common accidents that occur are cuts, burns, fires, poisoning and rarely, an explosion. Let us see one by one, what first-aid should be given to a victim when such a mishap occurs.

- Cuts: The most common accidents in the laboratory are cuts from broken glassware. If you have a cut, wash the wound well with cold water immediately. If bleeding is severe, apply pressure directly on to the wound to stop the bleeding. Then an antiseptic cream can be applied to the wound with a proper dressing.
- Burns: Burns generally caused by hot equipment can be treated as the cuts are treated, that is, wash the burn part with cold water for some time and then apply Burnol to it.
- Fire: A small fire in a beaker, caused by the vapours of an inflammable liquid, can be extinguished by covering it with a watch glass.

  If the clothes catch fire, one should lie on the floor and fire can be smothered by wrapping a blanket around the body.
- Poisoning: If one happens to swallow a poisonous chemical, plenty of water should be given if the person is conscious. For a corrosive poison, calcium hydroxide solution (lime water) should be given as soon as possible. An antidote should be given only in the case of non-corrosive poisons.
- Explosion: Sometimes a faulty technique during the experiment can lead to an explosion. You should work with highly oxidising or explosive chemicals only under strict supervision.

11. REMEDIES FOR A FEW CHEMICAL REAGENTS

Chemical	Neutralising wash
• Acids like HNO_3, H_2SO_4, HCl	$NaHCO_3$ or 2M ammonium carbonate (leaves no residue on clothes), then apply vaseline or a soothing cream.

(Contd...)

Chemical	Neutralising wash
• Alkalies, e.g., NaOH, KOH, etc.	1M acetic acid, then apply vaseline or a soothing cream.
• Bromine	2M ammonia, keep the affected part dipped in $NaHSO_3$ till bromine is washed off, then apply vaseline.
• Phenol	Ethanol and then hospital treatment.
• Sodium	Ethanol on a cotton wood pad.